MANAGEMENT, MACHINES AND METHODS IN CIVIL ENGINEERING

CONSTRUCTION MANAGEMENT AND ENGINEERING
Edited by John F. Peel Brahtz

MANAGEMENT, MACHINES AND METHODS IN CIVIL ENGINEERING

JOHN CHRISTIAN

Memorial University of Newfoundland

A Wiley-Interscience Publication

JOHN WILEY & SONS   New York • Chichester • Brisbane • Toronto

Library of Congress Cataloging in Publication Data:

Christian, John, 1938–
  Management, machines, and methods in civil
engineering.

  (Construction management and engineering,
ISSN 0193-9750)
  "A Wiley-Interscience publication."
  Bibliography: p.
  Includes index.

  1. Civil engineering. I. Title. II. Series.

TA145.C53   624   81-2434
ISBN 0-471-06334-7   AACR2

Printed in the United States of America

10 9 8 7 6 5 4 3 2 1

To Veronica, Simon, and Neil

SERIES PREFACE

Industry observers agree that most construction practitioners do not fully exploit the state of the art. We concur in this general observation. Further, we have acted by directing this series of works on Construction Management and Engineering to the continuing education and reference needs of today's practitioners.

Our design is inspired by the burgeoning technologies of systems engineering, modern management, information systems, and industrial engineering. We believe that the latest developments in these areas will serve to close the state of the art gap if they are astutely considered by management and knowledgeably applied in operations with personnel, equipment, and materials.

When considering the pressures and constraints of the world economic environment, we recognize an increasing trend toward large-scale operations and greater complexity in the construction product. To improve productivity and maintain acceptable performance standards, today's construction practitioner must broaden his concept of innovation and seek to achieve excellence through knowledgeable utilization of the resources. Therefore our focus is on skills and disciplines that support productivity, quality, and optimization in all aspects of the total facility acquisition process and at all levels of the management hierarchy.

We distinctly believe our perspective to be aligned with current trends and changes that portend the future of the construction industry. The books in this series should serve particularly well as textbooks at the graduate and senior undergraduate levels in a university construction curriculum or continuing education program.

JOHN F. PEEL BRAHTZ

La Jolla, California
February 1977

PREFACE

From the beginning of the nineteenth century, when civil engineers began to form professional institutions and societies, until today, revolutionary changes have occurred in management, machines, and methods in civil engineering. In about 160 years small hand tools, manual labor, and the use of horses have been replaced in most jobs by modern sophisticated machines, such as scrapers with microprocessor controlled automatic transmissions and vibrating compactors with miniprocessors that analyze signals and make it possible to compact fill material to a specified degree of compaction. More recently, as these changes have occurred, it has been necessary for university civil engineering departments to make major changes in their curriculum in response to the requirements of industry.

It is the intention of this book to help civil engineering students in their studies, particularly in courses that do not concentrate on analytical methods but which emphasize the practical and managerial aspects of civil engineering. It should also be a useful reference book to the young graduate civil engineer with a limited amount of on site experience.

An attempt has been made to use the industrial and academic experiences that I have gained to compare the differences in management, machines, and methods in civil engineering. Of the methods considered, comparisons are sometimes mentioned in the text, but other times only the overall preferred methods are described. It is hoped that this procedure will be advantageous to the student, especially today, since some areas of the construction industry, for example, the area serving the oil industry, are very much international in outlook.

It has been my pleasure to teach over 1000 civil engineering students in North America and Britain, and I should like to thank them for their thoughtful responses to my lectures on the topics covered in this book. Their responses provided essential feedback for the preparation of a book of this nature.

I should like to acknowledge the help that many colleagues have given me, particularly those in the construction area, such as Hira Ahuja, Lionel Baff, and Wally Campbell, and also the encouragement given by Dean R. T. Dempster and Dean C. D. diCenzo of the Memorial University of Newfoundland and Professor C. B. Wilby of the University of Bradford. My grateful thanks are also extended to Pat Gibson, Jackie Henderson, Janet Coffen, Ethel Pitt, Edwina Newhook, Marilyn Tiller, Brenda Walsh, and Brenda Knowles, who shared the task of typing the original manuscript, and to Terry Dyer and David Oliver, who prepared most of the figures.

I am indebted to the following companies and organizations which contributed to the textbook by supplying photographs or data:

Air Cushion Equipment Ltd., Southampton, England
Balfour Beatty Construction Company, Thornton Heath, England
Barber-Greene Company, Aurora, U.S.A.
Barber-Greene Ltd., Bury St. Edmunds, England
Blaw-Knox Construction Equipment Inc., Mattoon, U.S.A.
Blaw-Knox Ltd., Rochester, England
Bomag (GB) Ltd., Eynsford, England
British Monorail, Brighouse, England
Butter Cranes Ltd., Hillingdon, England
Coles Cranes Ltd., Uxbridge, England
CompAir Ltd., Camborne, England
Canadian Portland Cement Association, St. John's, Canada
Carl Zeiss Canada Ltd., Don Mills, Canada
Caterpillar Overseas SA, Geneva, Switzerland
Caterpillar Tractor Company, Peoria, U.S.A.
Cementation Ground Engineering Limited, Rickmansworth, England
CIRIA, London, England
Drott (J. I. Case/Tenneco Inc.), Racine, U.S.A.
Eagle Iron Works, Des Moines, U.S.A.
Edmund Nuttall Limited, London, England
Gadsden Scaffold Company Inc., Gadsden, U.S.A.
General Motors Scotland, Newhouse, Scotland
Green's Rollers, Clitheroe, England
H. Leverton and Company Limited, Gildersome, England
Hyster Company, Kewanee, U.S.A.

Hyster Europe Limited, Basingstoke, England

Ingersoll-Rand Rock Drill Division, London, England

International Harvestor Company, London, England

J. I. Case Company Limited, Stanningley, England

JCB (Canada) Excavators Limited, Burlington, Canada

JCB Sales Limited, Rochester, England

Koehring Canada Limited Crane and Excavator Group, Brantford, Canada

Koehring Crane and Excavator Group, Milwaukee, U.S.A.

Liebherr-GB Limited, Hatfield, England

Manitou (Site-Lift) Limited, Hampshire, England

Markham and Company Limited, Chesterfield, England

Millars Wellpoint International, Barnham, England

Newfoundland Tractor and Equipment, St. John's, Canada

Ontario Laser Rentals Limited, Bolton, Canada

Pipe Jacking Association, London, England

Priestman Brothers Limited, Hull, England

Ruston-Bucyrus Limited, Lincoln, England

Sauerman Brothers Inc., Bellwood, U.S.A.

SGB Limited, Mitcham, England

Sir Robert McAlpine and Sons Limited, London, England

Terex, G. M., Hudson, U.S.A.

The Robbins Company, Seattle, U.S.A.

Thwaites Engineering Company Limited, Leamington Spa, England

Thos. Smith and Sons (Rodley) Limited, Rodley, England

Universal Conveyors Company Limited, Leicester, England

Volvo Canada Limited, Bramalea, Canada

W. Lawrence and Son, Basildon, England

Wacker Canada Limited, Mississauga, Canada

Winget Limited, Rochester, England

Zokor Corp., Aurora, U.S.A.

The following publications particularly requested that the source of reprinted figures and photographs be quoted in full:

Proceedings of the Institution of Civil Engineers, Paper No. 7270S, for Figures 12.5 to 12.8

Proceedings of the Institution of Civil Engineers, Paper No. 7670, for Figure 12.11

Concrete Magazine, Concrete, April 1973, for Figure 14.3

I would be grateful if readers would inform me of any inconsistencies that they may find in the text of this first edition.

JOHN CHRISTIAN

St. John's, Newfoundland
June 1981

CONTENTS

IMPERIAL UNITS TO S.I. UNITS— CONVERSION FACTORS

Imperial Units	S.I. Units
1 in.	25.4 mm
1 ft	0.3048 m
1 yd	0.9144 m
1 mile	1.609 km
1 in.^{2}	645.16 mm^2
1 yd^2	0.836 m^2
1 yd^3	0.7646 m^3
1 US gal	3.78 liter
1 Imp. gal	4.55 liter
1 ft/sec	0.3048 m/sec
1 mph	1.609 km/hr
1 lb	0.4536 kg
1 lbf	4.448 N
1 ton (2240 lb)	1016.05 kg = 1.016 tonne
1 tonf (2240 lbf)	9.964 kN
1 ton (2000 lb)	907.18 kg = 0.907 tonne
1 tonf (2000 lbf)	8.896 kN
1 lbf/in.^{3}	157.1 N/m^3
1 lbf/in.^{2}	6.895 kN/m^2
1 lbf/ft^2	47.88 N/m^2
1 tonf (2240 lbf)/in.^{2}	15.44 MN/m^2
1 tonf (2000 lbf)/in.^{2}	13.79 MN/m^2

MANAGEMENT, MACHINES AND METHODS IN CIVIL ENGINEERING

CHAPTER 1

AN INTRODUCTION TO CONSTRUCTION MANAGEMENT

1.1 INTRODUCTION

Traditionally universities have concentrated on the analytical methods in the various branches of civil engineering. Today large numbers of graduates from civil engineering programs are employed by construction companies and a more practical knowledge of construction is more relevant to some graduates than the theoretical aspects of civil engineering. Accordingly, some universities have made major changes, in response to the requirements of the construction engineering industry, by instituting construction management and organization courses.[1]

A university civil engineering program should include the following topics:

Project planning

Programming

Management techniques

Construction organization

Construction practice

Communication

Contract documents

Legal problems in the construction industry

Equipment usage and acquisition

Temporary works design

Time-cost optimization

Cash flow forecasting

Computer techniques

Environmental studies

Architectural studies and aesthetics

The purpose of this textbook is to enable the young undergraduate or graduate civil engineer to understand the rudimentary principles of some of the abovementioned topics.

1.2 SOIL AND ROCK TYPES

When dealing with the management of equipment on construction sites the civil engineer needs to consider the soil or rock types that are present on the site. Since a considerable proportion of construction equipment is used to excavate, transport, and fill areas, it is very important to be able to identify all the soil and rock types present within the contract limits so that their properties may be determined or assessed.

It is often sufficient in equipment management to identify a particular soil type on site and assume certain properties that characterize that particular category of soil. The use of this type of assessment of the properties depends on the particular operation. If there is any doubt about the soil properties, it is the civil engineer's responsibility to decide if the operation warrants the laboratory testing of soil samples taken from the site.

In earthwork operations geological deposits are usually divided into two groups. Soils refer to relatively soft, loose, and uncemented deposits, often of comparatively recent geological origin. Rocks refer to hard, rigid, and well cemented deposits.

To characterize soil types for site identification purposes the size and nature of the particles and grains are classified. Each group is then subdivided, with the subgroups classified on the basis of particle grading.

A simple classification of soils for the purposes of site identification is:

Course grained, noncohesive soils

    Boulders

    Broken rock, hardcore

Gravels and sands

Gravel-sand, silt, clay mixtures
Well graded sands
Poorly graded sands
Sand, silt, clay mixtures

Fine grained, cohesive soils

Silts
Clays with low–medium plasticity
Clays with high plasticity

Organic soils

Peat
Other organic material and topsoil

More detailed descriptions of soils may be found in soil mechanics textbooks, standards, and codes of practices. To present a unified soil classification English language publications use the Casagrande classification.[2]

A description of the soil type found on a site should accurately give the type of particles, the density and structural characteristics, and the color. In a more detailed soil mechanics investigation a standard symbol based on the Casagrande group symbol helps further identify the soil but their use is unusual in the management of construction equipment on site.

The geological classification of rocks is very complex. Very detailed classifications are unnecessary in engineering but an ability to accurately describe rocks is essential. Accurate descriptions of rocks are required in discussions involving questions related to the blasting, hammering, ripping, or scraping of rock or soil deposits.

The classification of rocks is divided into three groups depending on their manner of origin, namely, igneous, sedimentary, and metamorphic rocks.

Magma originates below the earth's surface and solidifies, forming igneous rocks that are usually subdivided into rows and columns as shown in Table 1.1. In the columns the position of the rock depends on its acidity, or silica content. Acidic rocks are light in color and basic rocks are dark owing to the mineral content of the rocks. The position of the rock in the rows depends on the location of the rock when solidifying. When the extrusive rock came out of the earth's crust in magmatic form and cooled relatively quickly, the resulting rocks are

TABLE 1.1   Igneous Rocks

	Acidic	Intermediate	Basic
Extrusive	Rhyolite	Andesite	Basalt
Minor intrusive	Quartz-porphyry	Microdiorite	Dolerite
Major intrusive	Granite	Diorite	Gabbro

either very fine grained or noncrystalline. When the rocks cooled slowly within the earth's crust, such as the major intrusive rocks, the rocks are coarsely crystalline.

Sedimentary rocks were originally laid down from older rocks in sediments, by the action of water and wind, which then formed solid deposits of the earth's crust. Sedimentary rocks are subdivided by their manner of formation, namely, mechanically, organically, or chemically formed, as follows:

Mechanically formed

    Breccia

    Conglomerate

    Sandstones

    Shale

Organically formed

    Limestones

Chemically formed

    Limestones

The mechanically formed sedimentary rocks are subdivided according to the size of the particles, whereas the other two groups are usually subdivided according to the most dominant chemical character of the rock, although this method of subdivision is also used for sandstones in the mechanically formed group.

Metamorphic rocks are rocks that have been formed from other deposits by the action of heat or pressure or the combination of both. The main ones are:

Thermal metamorphic rocks

    Hornfels

    Quartzite

Stress metamorphic rocks

  Slate

  Schist

  Gneiss

1.3 LAYING OUT FOR CONSTRUCTION EQUIPMENT

Considerable thought should be given to laying out, or setting out, the lines, elevations, or levels, to control the work to be done by construction equipment. The length and width of the equipment should be considered to prevent dislocation of stakes or pegs. Access for other machines required to service the main equipment, such as trucks removing spoil, and the position of spoil heaps should also be considered.

In laying out excavation areas for footings, for example, the site engineer should use stakes for the center lines. Stakes should be painted white, easily visible, and clearly marked. Offset stakes may be necessary so that the site engineer can reestablish center lines after excavation at the bottom of the excavated area. The outside perimeter at the top of the excavated area must contain an allowance for the slope of the sides and be clearly obvious to the machine operator. Lines of sand, paint, or stakes may be used for this particular job. Extensive supervision is often required during this type of excavation. When the excavation work is near completion, stakes may be used to give the exact blinding level.

The center line should be staked out for the excavation of trenches. Stakes should normally be placed at a 25 to 30 m distance apart and at the center of each manhole. Offset stakes must also be positioned. Again, if it is feasible, a line of sand or paint placed on the center line, greatly assists the machine operator.

Figure 1.1 shows the use of profiles and travelers to give the depth of a trench excavation. Usually longitudinal profiles are used, but transverse profiles should be used when equipment or spoil heaps are likely to obstruct sighting in the longitudinal direction or when the sides of the trenches are sloped so that accurate sighting becomes difficult. At all times the width of the machine to be used should be considered in the positioning of profiles. Important stakes should be fixed securely or concreted in. Profiles should be sturdy and the cross piece on the traveler should be stiffened by a gusset or tie piece. As the excavation nears the required depth, another system of leveling within the trench,

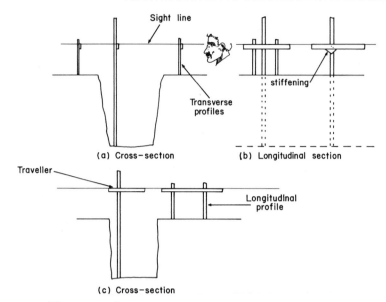

Figure 1.1   Laying out excavation elevations for trenches.

using stakes or profiles, may be necessary to achieve sufficient accuracy. The use of self-leveling laser beam equipment is gaining popularity for large scale operations of this type.

The laying out procedure for constructing embankments is shown in Figure 1.2. The earthwork is constructed in layers so that a satisfactory degree of compaction may be achieved. Benching and overfilling by at least 150 mm at the side slopes is necessary so that the edge of the embankment is thoroughly compacted by compaction equipment. The formation is then trimmed to the final elevation required. Batter board profiles, as shown in Figure 1.2, are suitable for controlling elevations. Separate travelers can be made for the overfill elevation, final elevation, and topsoil elevation.

For high slopes further profiles are necessary for the accurate control of the embankment. The final formation level should be temporarily overfilled to a minimum depth of 150 mm if constructional equipment is to be used on the embankment.

At all times care should be exercised to position profiles or batter boards so that the banksman can easily sight along the cross or batter pieces, and frequent checks should be made to see if there has been any movement due to stakes or profiles being inadvertently knocked by site equipment or personnel.

Plate 1.1 Laser alignment. Courtesy Ontario Laser Rentals Ltd. Bolton, Ont., Canada.

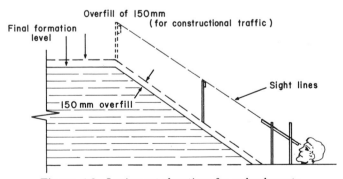

Figure 1.2 Laying out elevations for embankments.

7

1.4 OWNER, ENGINEER, AND CONTRACTOR RELATIONSHIPS

A person or group, known as the owner, may decide to initiate and promote a project. The owner may be alternatively referred to as the client, employer, or promoter. The owner is responsible for providing money and all information for execution of the project. The owner can be federal, state, or provincial governments, city or metropolitan district authorities, an entrepreneur, hydro, telephone, light and power companies, or air or rail organizations. Often the owner is not conversant with the theoretical and practical techniques used in civil engineering and therefore employs an independent consulting civil engineer, often referred to as the engineer, to advise him on the civil engineering aspects. In building and some types of construction work an architect instead of an engineer may be employed by the owner.

The engineer's responsibility is to consider the feasibility of the project and to prepare a safe design at the lowest cost. The engineer then provides contractors, who intend to bid, or tender, for the execution of the works or project, with a set of contract documents consisting of the bid or tender form, the form of agreement, drawings, the specification, the conditions of contract, and often a bill of quantities, depending on the type of contract. The engineer, or architect, may often be given the responsibility to supervise the construction of the project and issue interim and final certificates for payment of the work completed. The owner, engineer, and contractor can appoint representatives or assistants to conduct negotiations before and during construction of the project. Generally the engineer recommends that the contract be awarded to the lowest bidder. In some cases this is mandatory unless there are very special reasons that the contract should not be awarded to the contractor who has submitted the lowest tender. Exceptions might include very specialized work, obvious errors, or doubts about the contractor being capable of constructing the projects for any reason, such as the sheer size of the project.

The owner enters into two contracts. One contract is between the owner and a consulting civil engineer, and the other contract is between the owner and a contractor. In both contracts the duties and responsibilities of the engineer, contractor, and owner toward each other are stated. The contract between the owner and the contractor states the method and terms of payment for construction.

In the event that the contractor is unable to carry out the work according to the contract documents the contractor may be required to undertake to pay money for an incomplete contract. This is known as a surety. Alternatively, the contractor can obtain the guarantee of an insurance company or bank. The terms of this agreement or bond are usually quite specific but vary for different authorities, different areas, and different types of projects. A person or organization who is responsible for the default or debt of the contractor is also referred to as a surety.

1.5 BIDDING

The bidding or tendering practice is common in most civil engineering projects. Usually the bids are opened on a designated date. Occasionally a contract is awarded to a selected contractor due to the unusual, specialized, or secret nature of the project.

The profit or loss on a project partly depends on the bid itself. An estimating department must have a good knowledge of the current costs in construction and make allowances for uncertainties such as inflation. There should be a very clear demarcation between contracts that are index linked and those that are not. If the total bid price is too high, the contractor loses the contract, but if the total price is too low, the contract may be won even though it is likely to make a loss. To obtain a good balance the contractor is usually influenced by the amount of risk, the prevailing economic position of the company and the industry, and the size of the project. These factors are often sufficient to win or lose a contract and ultimately determine whether a successfully bid project makes a profit or loss.

The probability of the success of a tender obviously depends on the number of companies bidding for a particular project.[3] A diagram of this effect is shown in Figure 1.3.[3] The likelihood of success increases the more the contractor's estimate is less than the engineer's estimate; this is also shown in the diagram.

The effect of the economic climate may be seen in Figure 1.4.[3] Over a

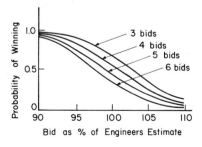

Figure 1.3 Probability of winning bids versus engineer's estimate.

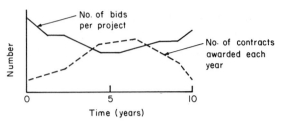

Figure 1.4 Number of new contracts per year compared with the number of bids per project versus time.

certain period of years the average number of contractors bidding for similar types of projects varies. When the economic climate is good, more projects are started in a particular year, and the result is that the number of contractors bidding for each project decreases.

When the economic climate is good, the actual average bid as a percentage of the engineer's estimate also increases. This therefore reflects market forces and has an inflationary effect on the construction industry. These factors demonstrate the very great difficulties and uncertainties for the contractor in estimating for a civil engineering project. The decision whether to bid for a particular project also requires careful examination since the operation of bidding alone is expensive to the contractor.

It has been mentioned previously that occasionally the contractor who has submitted the lowest bid is not awarded the contract. If a contract is awarded to a company that is not the lowest bidder, the owner or his representatives usually take extra precautions to ensure that the special reason for selection of the contractor is fulfilled.

In civil engineering the main contractor who is awarded the contract generally subcontracts certain sections of work to other contractors. For a particular section of work the main contractor will have already received estimates from several subcontractors. These estimates in fact are generally very closely scrutinized by the main contractor in the formation of the main bid. After postbid negotiations the main contractor either decides to carry out a section of work alone or offers the section to a subcontractor who has submitted the lowest estimate. This estimate may be revised during postbid negotiations. Some of the subcontractors may reexamine their wish to participate in the section of the contract because of a changed work load situation in the time that has elapsed since the original estimate.

1.6 METHODS OF BIDDING

There are several methods of managing the bidding and contracting stages of a civil engineering project. A negotiated contract is one where the owner or promoter negotiates with a contractor to build a project for a particular price or at a particular rate. The contractor or builder may purchase land, design, develop, and construct the project. The method can be used, for example, in the manufacture of prefabricated houses and the building and equipping of apartment blocks at a negotiated rate per square meter of floor area. The disadvantage of this type of contract is the lack of competition and the insensitivity to rapidly altering market conditions. Public works projects usually require that a contract bid should be open to competition.

The most frequently used method involves bids that are issued in open competition. Sometimes contractors are invited to bid according to their previous experience, but generally any contractor is permitted to bid for a civil engineering project after it has been advertised. The lowest bidder is then awarded the contract. The contractors usually limit their own bids according to the size and type of the project.

Another recent method of managing the bidding and contracting stages of a civil engineering project is by the appointment of a project manager. The project manager who is hired occupies a position similar to that of the construction manager of a main contractor except that the project manager oversees the entire project across company or departmental lines. Project management provides a team approach and is successful on very large projects where various parts of the project are separated into stages and trades, and tenders are invited, by the project manager, for the individual stages in sequence.

1.7 TYPES OF CONTRACTS

In the traditional contracting method used in civil engineering all the bids are based on the contract documents, and a fixed or stipulated price for the contract is submitted to the engineer from each contractor. The total cost of the project is therefore known when the contract is awarded to the contractor who has submitted the lowest bid, except for the additional cost of unforeseen site conditions or design changes. In times of high inflation an index-linked contract price fluctuation clause may be included in the conditions of contract.

In a unit price contract each unit of work is itemized in a bill of quantities. The quantities are listed in the bill for each item and the contractor, after very careful consideration, enters a price for each unit. The sum of the amounts for all the items then gives the contractor his total amount to submit as his bid. Table 1.2 shows a typical item in a bill of quantities. Usually there are general items at the beginning of the bill of quantities which do not possess a unit rate but are presented as a sum.

On completion of the project the actual quantities may vary somewhat from those entered in the bill of quantities. Providing that the variation in quantities is not considerable, the contractor is only paid for the actual quantities. It is therefore the rates that are fixed, not the total amount of the contract value.

The unit price contract has several advantages. An equitable method is produced during the tender stage, as each contractor bids on exactly the same quantities and items. The quantities and details of work are clearly listed, and this simplifies payments for alterations to constructional details during the project. It is also simpler to estimate the amount of work currently executed for each interim monthly payment.

Sometimes the work is not itemized in a bill of quantities, and after competitively bidding the successful contractor agrees to undertake the project specified in the contract at a price or lump sum agreed to and fixed or stipulated at the time the contract is signed. This method is particularly appropriate for a project without any major alterations or where unforeseen circumstances are extremely unlikely, such as the repetition or extension of an existing job. Although the contractor will have itemized the work in the project at the tender stage for the purpose of obtaining an accurate estimate, detailed measurements and accounting are eliminated as the project progresses.

The main characteristic of the fixed price type of contract is that the contractor is under a contractual obligation to carry out the work at a fixed price. The owner therefore knows the price to be paid on completion of the project providing that there have been no alterations.

There are situations where a cost plus fixed fee type of contract may be used. The owner may wish to start a project without finalizing all

TABLE 1.2 A Typical Item in a Bill of Quantities

Item no.	Description	Quantity	Unit	Rate	Amount
607	Reinforced concrete fcu = 40	10	m^3	70.00	700.00

the design details. Alternatively, the contractor may not want to risk a fixed price type of contract because of the uncertainties or size of the job or because of local conditions prevailing at that particular time. The contractor therefore agrees to carry out the work at actual cost, with an additional fixed fee to cover management, overheads, and profit. The on site costs for all items are therefore reimbursed by the owner. The fixed fee is either for an agreed fixed amount or an agreed fixed percentage of the actual costs. The percentage usually decreases as the cost of a project increases. There is less risk to the contractor in this type of contract provided that the duration of the project can be estimated fairly accurately. There is perhaps more risk to the owner, however, because there is less pressure on the contractor to produce the most economical job.

Perhaps the most economical overall type of contract is the package deal, alternatively referred to as the turnkey or all-in type of contract. In this type of contract the contractor undertakes to design and construct the project to completion. The disadvantages are that if two or more contractors have been approached, the designs may be completely different, making comparisons difficult. Also, the system does not provide an unbiased party to settle disputes. The project management method of contracting may differ very little from this type of contract.

Finally, an authority may retain its own direct labor department to carry out construction work. In this case the owner or president employs an engineer to design and supervise work carried out by the authority's own direct labor force. A contract is generally not used for this type of work unless it is interdepartmental, but there is usually some method of internal accounting. It is a useful system for specialized work but is often very uneconomical because of work load fluctuations.

Sometimes a clause is written into the contract documents which enables the successful contractor to reassess the method of construction, and even the design and use of materials, so that he can suggest alternatives that would produce substantial savings in the contract price. If the reassessment indicates to the contractor that savings can be made, a document is formally presented to the owner in which the alternatives are fully explained. If the owner agrees to the alternatives, the savings are split between the owner and contractor. There are valid reasons why a contractor may be able to suggest cheaper alternatives. For example, a modification to the design may mean that equipment used on a previous project may be used after adaptation.

This process of reassessment is sometimes referred to as value engineering. The owner may ask the consulting engineer to make a reassessment before the contract is awarded. The consultant may engage the services of a contractor during this stage. A reassessment can be made either before or after a contract is awarded.

1.8 INTERIM PAYMENTS

At the end of each month the contractor usually submits a statement to the engineer so that an interim payment can be made for the work carried out and materials supplied during that month. The statement should show the estimated contract value of the permanent works carried out, the value of any materials delivered to the site but not by then incorporated in the permanent works, the value of any materials not delivered to site which have been paid for, and the estimated amount to which the contractor considers himself entitled in connection with temporary works and constructional equipment for that month.

Example of Monthly Statement

(a)	Permanent works	$285,000.00
(b)	Materials delivered to site	90,000.00
(c)	Temporary works and equipment	125,000.00
		500,000.00
	Less 3% retention	15,000.00
	Interim payment	$485,000.00

It can be seen in the simplified example of a monthly statement that a certain percentage is retained by the engineer. This percentage may vary from 3 to 10% depending on the value of the project.

The engineer should certify that the contractor's monthly statement is correct and the owner should then pay the amount within one month. The retention money covers the owner for any contractor's errors, defects, or withdrawal. Half of the retention money is repaid after the certificate of completion is issued by the engineer, and the other half is returned after the expiration of the period of maintenance.

1.9 PROJECT FINANCING

The cost of a one million dollar project against time is shown in Figure 1.5. The expenditure by the contractor describes a typical S-shaped

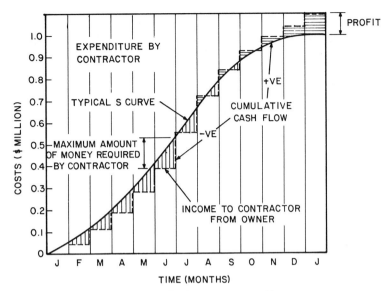

Figure 1.5 Project costs versus time.

curve. At the end of each month the contractor receives an interim monthly payment for the work that has been carried out during that month. These payments are shown by broken stepped lines in Figure 1.5. If a one million dollar project is considered, with a project duration of 12 months, the maximum amount of money required by the contractor to finance the contract is the maximum negative value between the stepped income line and the expenditure curve.

The net cumulative cash flow is the hatched area in the figure. The vertical hatching represents the negative cumulative cash flow, which is the amount of money that the contractor requires to finance the project. The horizontal hatching represents the positive cumulative cash flow, which represents the amount of profit that the contractor eventually makes, at the end of the contract, that is, about $100,000. The amount of retention money is omitted to simplify the graph. The profit can be used to pay shareholders, pay off interest, finance subsequent projects, and increase the assets of the company.

It can be seen in Figure 1.5 that the maximum amount of money required by the contractor to finance the project is about $130,000. The provision of this finance is obviously an extremely important function of the board of directors. The method of obtaining this finance tends to vary and depends on the size of the company.

For smaller companies it is necessary to approach a bank and establish credit for use on one particular project or, alternatively, a specific amount of credit for use on a number of projects. The bank investigates the company's recent financial statements and, if the request is successful, proposes the terms and method of repayment, including interest. If the contractor is well known to the bank and has established credit worthiness over a number of years, the procedure is obviously more straightforward and requests are likely to be granted more easily.

The interest rates charged by the bank usually vary. The interest rate is likely to be at the same level or higher than the prime lending rate, or bank rate. The bank considers the assets and liabilities of the company, reviews its previous experience with the company, and, from the financial statements, assesses the company's likelihood of making a profit. The bank's own availability of finance and the current economic position of the country also influence the decision. Sometimes another large company or individual may back the small contracting company by providing credit for a project.

After arranging the finances, the contractor then arranges a bond from an insurance company or bank. If sureties or performance bonds are required, the contractor needs to approach an insurance agent, who may need to approach one or more insurance companies as sureties. The companies issuing the performance bond guarantee to the owner that the project will be completed, in the specified time and at the tendered price, and that any financial loss that the owner might incur, due to the contractor's failure to complete the project, will be made good. A person or company that guarantees to make good any default by the contractor is called the guarantor or surety. Once the surety companies have verified that the contractor has sufficient finances to carry out the project, it issues a bid bond, which is presented with the contractor's tender. This is a statement that the companies expect to write the bond if the contractor is successful with his tender and guarantees that the contractor will enter into a formal contract to construct the works. A deposit is sometimes used as an alternative to a bid bond.

Large established contracting companies may start large projects every month or so. For this reason they may establish credit facilities with several banks so that the credit may be arranged immediately. For very large projects two or more contractors may set up a joint venture so that financing arrangements as well as resources may be pooled for this particular project.

1.10 PREPARATIONS AND PRELIMINARIES FOR CONSTRUCTION

Many activities must start even before the contract has been awarded to a contractor.

Permits to commence construction of the project are necessary. The design engineer or architect will probably be required to make a presentation to the local planning board. The environmental effects and impact should be described accurately and shown to interested parties. This is a very important and essential preliminary to any scheme so that the necessary modifications arising from valid and sensible objections can be incorporated in the scheme, thereby avoiding long delays, which generally substantially increase the overall cost of a scheme. Environmental impact studies may be necessary.

The consulting engineer has to prepare the contract documents. These documents include the form of tender, the specifications, the conditions of contract, the contract drawings, and the form of agreement and often also include a bill of quantities. The drawings alone may take many months to prepare. Site investigations into the soil types and conditions existing on site are usually necessary. Often very extensive site investigations are necessary to determine the soil conditions so that the foundations of a structure can be designed safely and economically.

A preliminary layout survey is necessary so that up-to-date and accurate details within the contract limits can be shown on the contract drawings. Sometimes it is only necessary to verify the information shown on existing plans or maps. Certain location points should be shown on the drawings to enable the contractor to lay out the relevant center lines and elevations of the foundations, earthworks, and pipework.

It may be necessary for the contractor to make preliminary inquiries and conduct tentative negotiations for the supply of some materials during the tender stage. If, for example, large quantities of rock need to be imported for fill, the contractor must investigate local quarries or make arrangements with landowners for the establishment of a borrow pit, if only to give an accurate rate to the chief estimator.

When a contractor is awarded the contract, it may be necessary to employ a demolition subcontractor immediately to demolish some existing property. The main contractor should be particularly concerned with safety during this operation and should ensure that a reputable and experienced company carry out the demolition.

Plate 1.2 Electronic angle and distance measurement. Courtesy Carl Zeiss Canada Ltd. Don Mills, Ont., Canada.

Before any work is commenced on site, however, various types of insurance are required to protect employees on site, members of the public, and adjacent property.

1.11 HAUL ROADS

Sites covering large areas, such as open-pit mining, quarrying, hydro, and highway schemes, require a well planned system of haul roads for removing excavated material, importing fill, and supplying many

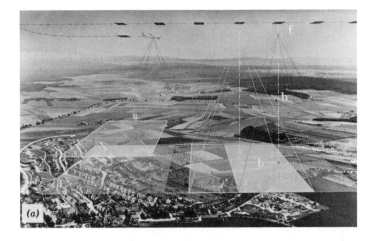

Plate 1.3 Aerial survey. Courtesy Carl Zeiss Canada Ltd. and Don Mills, Ont., Canada.

types of materials. Access to site is normally described in the contract documents. The type and location of a haul road depend on the distribution and location of the many sections of the civil engineering works and the type of equipment to be used. Haul roads should be located where the elevation is not changing frequently due to cut and fill activities.

Usually haul roads for access should be made with a thickness of 200 to 500 mm of hardcore, the depth depending on the type and condition of the subgrade. An initial expenditure on a good quality, thick haul road is recouped quite quickly compared with a poorly made, badly maintained haul road. When heavy equipment uses haul roads, rain and snow can soon disrupt and delay traffic. Maintenance is essential. On large sites the use of a motor grader to continuously maintain haul roads can be justified economically. The amount of maintenance depends on the type of soil. Obviously a soft, muddy soil requires more maintenance than harder rocks.

On very long haul roads where the traffic consists mainly of trucks hauling material, a proper metaled road can be economically justified, particularly if the road is required for several years. However, the use of tracked machines, particularly tracks with grips, on a metaled road soon ruins the surface and the road quickly disintegrates.

Mesh blankets provide a satisfactory haul road and are useful for immediate or temporary use. The blankets can be recovered and are reusable. Concrete rafts or railway sleepers are sometimes used to distribute wheel loads if heavy traffic is required to cross over shallow underground services.

1.12 SITE FABRICATION

It is necessary to order many materials for the construction of the project as soon as the contractor has been informed that his tender has been successful. As soon as the site office has been set up and tradespeople have been recruited, work on some materials can commence. Close scrutiny of the critical path program enables a decision to be made on whether immediate work is either necessary or economical.

Steel reinforcement may have been delivered to site in straight lengths. Usually bars of diameter between 10 and 50 mm are delivered in about 13 m lengths, while smaller diameter bars are delivered in shorter lengths. The steel can be cut and bent according to the bending schedules and stacked in the site compound ready for use in the reinforced concrete foundations. Alternatively, it may be decided to have the steel reinforcement cut and bent before delivery to site.

Structural steelwork is usually prefabricated in a workshop and delivered to site. Only bolting and small areas of site welding are therefore necessary on site. Prestressed concrete units are also prefabricated in a workshop. Most units are pretensioned and little work is required on site, but for post-tensioned units the amount of work on site can be quite elaborate. There is a trend these days to prefabricate as much work as possible in a workshop, where quality control is more reliable and the effect of weather on production is insignificant.

Sheets of plywood and lengths of timber that are delivered to site are usually prefabricated into modular sizes for formwork. Further information on the prefabrication of formwork may be found in Chapter 11.

1.13 WORKS LAYOUT

Often on small construction sites, particularly in the centers of towns and cities, there is very little space available at ground level for the office, plant yard, and works. In busy streets it is often necessary to have the site office temporarily located on scaffolding at an elevation above street level. Stores and stacks may also be necessary at the office elevation. However, on large civil engineering projects there is usually plenty of space available for the works layout. A typical works layout is shown in Figure 1.6.

The contractor's main offices should be near the main access point if possible. The checker's cabin and stores should be close to the main offices. The resident engineer decides the location of the office for his staff but this should preferably be close to the contractor's main office. The proximity of these offices helps create an efficient means of communication. A contract that continues for two or three years may have a peak period during one summer only. It is therefore often useful to extend some of the offices during this peak period only. Provision should therefore be made for the extension of some of the offices.

A fairly large clear area should be allocated for the cutting, bending, and stacking of reinforcement. The rebar workers' work will be made easier if a straight production line is created. The steel reinforcement should be stacked above ground level. Money is often wasted on the cleaning of mud splashed reinforcement. Stacking and marking of reinforcement should also be carefully planned. Time is often wasted searching for and removing reinforcement from stacks.

1.14 TEMPORARY WORKS AND SERVICES

Temporary works are the works that are required for the completion or maintenance of the project but are subsequently removed because these works do not form part of the permanent works. In most cases the contractor designs the temporary works. The contractor should always submit drawings and design calculations for important temporary works to the engineer, who should check and scrutinize them carefully.

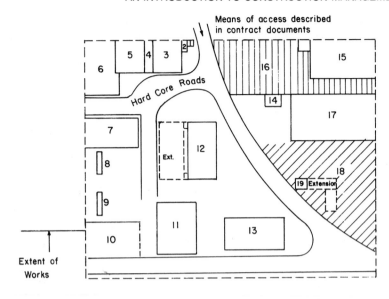

1. Checkers cabin
2. Storekeeper
3. Stores
4. Drying room
5. Mess
6. Canteen
7. Steel stack
8. Cutting m/c bench
9. Bending
10. Steel stocks
11. Carpenters shop
12. Contractors offices
13. R.E. offices
14. Concrete testing lab.
15. Fitters shop
16. Hard standing – plant parking
17. Batching plant
18. Hard core or concrete base
19. Sub–contractors

Figure 1.6 Works layout.

This check, however, does not relieve the contractor of his contractual responsibilities for the safety of the temporary works.

The provision of temporary services is required as soon as the contractor arrives on site at the beginning of the project. Apart from the provision of huts and offices the contractors, engineers, and subcontractors require a telephone service. As the telephone is so essential for communications, the project manager should arrange the installation of telephones with the local telephone manager immediately. Problems may arise in remote areas where the nearest telephone line may be some distance from the site area. Lines may be erected overhead from one office to another unless equipment with high headroom is likely to cross the line, in which case heavily reinforced underground ducts should be used.

The water supply connection to the water main is usually made in a trench by the plumbing subcontractor or a local plumber hired for that particular job. The pipe diameters should be designed according to the estimated peak consumption. Water is required for drinking and industrial use and may also be required for fire protection. Connection to a water main may be impossible in remote areas.

An electricity supply is required for lighting, heating, and small power tools. The local power supplier should be contacted immediately to arrange a temporary connection to the site. The contractor's electrician can then make the underground connection to the offices and huts. In remote areas where the nearest power supply is a long distance from the site, a generator is required. Small civil engineering tools sometimes require a transformer.

As soon as employees work on site, sewerage and sanitary facilities must be provided. Initially, portable chemical toilets may be provided, but if the project is to last any length of time flush toilets should be provided. Washing and toilet facilities should be designed for the peak number of personnel during a project. The local public health authority should be consulted regarding the disposal of effluent. Effluent may be disposed of by connecting the pipes on site into the mains, making use of a holding tank, or building a small scale sewage treatment plant.

Heat can be provided by electricity, oil, or propane gas. Fire regulations have become more stringent, and care should be taken that the storage of fuel tanks or drums conforms to the latest fire and safety regulations.

If services are to be diverted, the project manager must arrange meetings with the various statutory authorities. If a resident engineer is present on site, joint action should be taken because the resident engineer's design office will have already had discussions on the location of all the services with the various authorities. Determination of the exact location of existing services on site frequently causes difficulties. Many authorities prefer to divert and permanently reinstate service lines themselves.

CHAPTER 2

SCOPE OF CONSTRUCTION ORGANIZATION

2.1 WORK ORGANIZATION AND CLASSIFICATION

Work should be organized into groups of from 4 to 10 persons. This size of group can be managed effectively, and individual or group responsibilities can be clearly defined. Individual and group productivity per individual is reduced as a group becomes larger.

Organizational planning establishes the links between individuals and groups. Communication within a group and between groups is very important. The functions of each group should therefore be defined clearly so that there can be efficient coordination and communication between each group supervisor. Sometimes each group may perform relatively independent tasks that require careful monitoring by the supervisor and close liaison between other group supervisors to ensure that the job as a whole is progressing in a coherent and coordinated way.

When a small company expands, it is necessary for the owner to delegate responsibility and institute some organized lines of communication within the company. For large civil engineering companies a good organization chart is essential. An organization chart is a graphic representation of the organization showing the chain of command and the functions of each individual, group, or department. Often more than one organization chart is necessary. The head office organization of a contractor's office is different and separate from the organization of each individual site office although there should be frequent interdisciplinary liaison between the head office and site office. Newly recruited employees should be shown the organization charts during their training period and should also be provided with the chance of working in various departments during their development within the company.

In civil engineering, organization charts are produced showing the contractor's and consulting engineer's organization both at the head office and on site. Typical charts are shown in Sections 2.2 to 2.5.

The types of work frequently encountered in civil engineering should also be classified into groups[4] so that a framework can be used in the organization of the company and for measurement in contract control.

A work classification similar to the following should be used:

Class

A	General items
B	Site investigation
C	Geotechnical and other specialist processes
D	Demolition and site clearance
E	Earthworks
F	In situ concrete
G	Concrete ancillaries
H	Precast concrete
I	Pipework—pipes
J	Pipework—fittings and valves
K	Pipework—manholes and pipework ancillaries
L	Pipework—supports and protection, ancillaries to laying and excavation
M	Structural metalwork
N	Miscellaneous metalwork
O	Timber
P	Piles
Q	Piling ancillaries
R	Roads and paving
S	Rail track
T	Tunnels
U	Brickwork, blockwork, and masonry
V	Painting
W	Waterproofing
X	Miscellaneous work

2.2 CONTRACTOR'S HEAD OFFICE ORGANIZATION

Large companies contain a board of directors, with a chairman of the board directing all operations. Sometimes the companies possess a vertical type of organization, while in other companies a functional type of organization has developed with some horizontal links, similar to the one shown in the bottom part of Figure 2.1.

The head office organization of different contractors' companies therefore varies. The development, size, and degree of specialization of the company all tend to produce variations in company organizations. However, the lists in Figure 2.1 show the typical functions that need to be carried out within a general civil engineering contractor's head office organization.

Often regional offices need to be established. These offices may be managed by one or more of the directors. There are several advantages

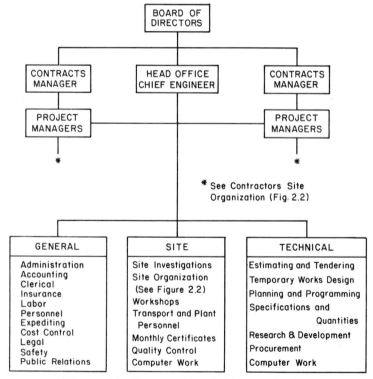

Figure 2.1 Contractor's head office organization.

to a regional office. A plant yard may be set up which reduces the traveling distances for equipment and mechanics in the event of major breakdowns on site. The procurement of new projects is often easier when offices are spread out on a regional basis; local experience is important in establishing the cost and availability of materials, and the dislocation and relocation of key personnel are minimized, resulting in a more stable work force. The district manager may direct several projects from a regional office which generally contains small engineering and administrative departments.

2.3 CONTRACTOR'S SITE OFFICE ORGANIZATION

A large contractor's site organization is shown in Figure 2.2. It is a typical site organization for, say, a 25 million dollar highway project. There are considerable variations depending on the size, type, and degree of specialization of the project. For smaller jobs there are considerably less engineers on site and several job functions may be carried out by one engineer. Sometimes mechanical, electrical, or chemical engineers are required in the organization even though the job is mainly a civil engineering job. Other times surveyors may replace the section or assistant engineers. Job titles vary from one company to another and according to fashion.

The typical duties and responsibilities of each job title are discussed in Section 2.7.

2.4 ENGINEER'S HEAD OFFICE ORGANIZATION

As a consulting civil engineer's practice grows from a small group of partners, it is necessary to plan some organizational structure. Most large firms have an organization similar to the one shown in Figure 2.3. Over the last decade most offices have installed or arranged to share some computer facilities. This trend will continue as the comparative cost of work done by computer decreases. The number of section engineers or project engineers varies with the size of the firm and the number of different projects undertaken at a particular time. The section engineers frequently manage more than one project, particularly when the projects are small or when intermittent work is required on a project.

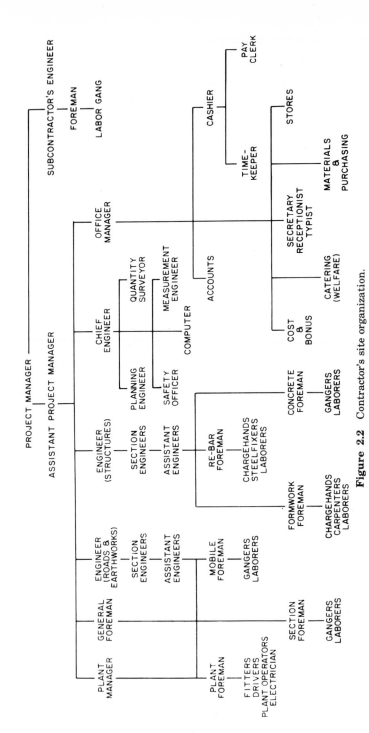

Figure 2.2 Contractor's site organization.

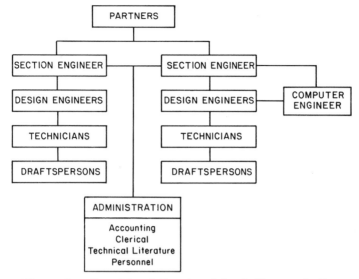

Figure 2.3 Consulting civil engineer's head office organization.

Firms often have regional offices with a partner or regional manager directing each office. The regional office generally has an organizational structure similar to that of the head office but on a smaller scale. In overseas work it is useful to have some local personnel who are familiar with the local conditions and contracting system to alleviate any difficulties arising during the planning, design, and construction stages.

2.5 RESIDENT ENGINEER'S SITE OFFICE ORGANIZATION

A typical resident engineer's site office organization is shown in Figure 2.4. These personnel may be employed by a consulting civil engineer's firm or a local, regional, or federal government's engineering office. On a large job an assistant resident engineer and two or three assistant engineers may be employed. On small projects only one qualified engineer may be resident on site. When the engineering and technical aspects are not significant, an inspector or clerk of works may replace the resident engineer's site office organization. The duties and responsibilities of personnel in a resident engineer's site office are discussed in Section 2.8.

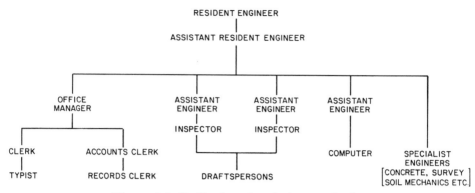

Figure 2.4   Resident's engineer's site organization.

2.6 PROJECT MANAGEMENT

The increase in size and complexity of some civil engineering projects
has made it more difficult to complete a major project within the limited time required and within the owner's estimated budget. For this
reason the owner requires an efficient and sophisticated management
process. A new process, called project management, has evolved in an
attempt to achieve greater efficiency, liaison, and sophistication. The
meaning of project management, in this narrower sense, is the system
of coordinating the planning, design, and construction of a project
within the time and budget required by the owner. In this type of
system the owner appoints a project manager to fulfill these requirements. This is not a new concept except for the fact that the project
manager may be an individual who is not necessarily connected to the
design engineer's staff or the main contractor's staff. Unfortunately,
the words "project manager" and "project management" are now am-
biguous, depending on whether the words refer to this narrower sense
or the general sense.

To explain the system in more detail it is useful to consider the
process of managing a civil engineering project in general terms.

The traditional system is shown in Figure 2.5a. The design engineers, irrespective of whether they supervise on site, are separate
from the main contractor. The main contractor can then decide
whether to employ its own labor or several subcontractors. Sometimes
a subcontractor is nominated by the engineer to carry out some particular type of work. Occasionally the main contractor may subcontract
all the work and act as a project manager.

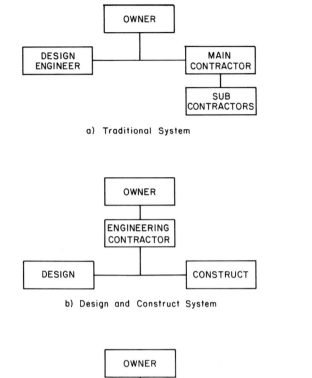

a) Traditional System

b) Design and Construct System

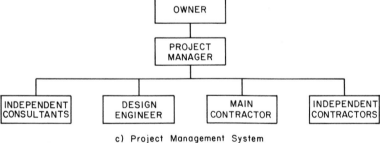

c) Project Management System

Figure 2.5 Process of managing civil engineering projects.

In the design and construct system a single contractor carries out the design and construction. This contract is usually operated on a lump sum or cost plus basis. Again, the engineering contractor may decide to subcontract part of the construction work. This system is shown in Figure 2.5b.

For very complex projects a system has evolved where a project manager is appointed to employ separate firms for the design and construction of the project and independently employ other consultants or contractors. This system is shown in Figure 2.5c. A professional fee is generally negotiated with the consultants and design engineers, and the contractors compete for fixed price or cost plus contracts. These contracts are let by the project manager on behalf of the owner. There are variations to this system. The main contractor may act as a construction manager and employ other contractors, or the consulting engineer undertaking the design may employ other specialist consultants. Alternatively the consulting engineer or even, occasionally, the main contractor may act as the project manager.

2.7   THE CONTRACTOR'S DUTIES AND RESPONSIBILITIES

Personnel at the contractor's head office are responsible for all the company functions and in a general way are responsible for all the projects. If there is a network of regional offices, the regional office takes over the general responsibility for the projects within that region. Overall company policies, however, emanate from the head office.

Company policies and procedures are approved by the board of directors. A director or vice-president usually supervises the execution of projects and coordinates the work of various departments, within the head office organization, with the contracts manager who is dealing with various projects at site level. In large companies a chief engineer assists the director in the development of construction methods, feasibility studies, and estimating.

The contracts manager or regional engineer supervises the contracts within his area or regional office. The contracts manager directly supervises the estimates, temporary works design, plant allocation, the development of construction methods, and the provision of key personnel on each project. This person should be able to provide technical and marketing advice to the project manager and should be able to provide specialized technical assistance and extra temporary personnel when required. Usually the head office has a small personnel office. The head of this office selects and trains both the engineering and administrative personnel.

The project manager or agent is the chief executive on site whose position is mainly managerial. A person in this position directs and controls the whole construction work on site, and only major decisions and policies are reported to the contracts manager. The project manager employs local nonprofessional personnel and directly negotiates with subcontractors, as well as hiring equipment, ordering, and purchasing materials. Obviously on large jobs some of these functions must be delegated. It is often necessary to employ an assistant project manager to assume some of the project manager's responsibilities. These responsibilities are further delegated to the plant manager, the general foreman, and the engineers responsible for a section of the project. The field engineers have direct control of the tradesmen and equipment in their particular area. Coordination between the planning engineer, general foreman, plant foreman, and the field engineers is essential for the supply of equipment and the movement of site personnel from one section to another. The field engineer assumes the day-to-day responsibilities for one particular area including materials orders, measurement, and the overall engineering aspects of the section.

An essential link between site management and the trade foremen is the general foreman's role. The general foreman directs the daily distribution of labor and equipment in consultation with the field engineers, foremen, and planning engineer. Frequent liaison with all parties is essential to produce a coordinated team effort. In practice this is one of the most difficult areas of site management. A critical path program may often alleviate difficulties in clearly showing the most urgent jobs.

The chief engineer is responsible for planning, quantity surveying, measurements collation, safety, and the computer, whereas the office manager is responsible for the nontechnical jobs such as the accounts, materials and purchasing, clerical staff, welfare, and wages.

The field engineer who is responsible for the structures on a particular project, for example, may have under his jurisdiction several section engineers, each one being responsible for an individual structure under construction on the project. The section engineers are responsible for layout surveys, programs and progress, temporary work schemes, and formwork design. They should keep detailed and accurate diaries to record progress and should be prepared to give immediate technical guidance to foremen. Frequently there are other less experienced engineers to assist the section engineers but this depends entirely on the work load. In countries where layout lines have a legal significance a licensed surveyor should be employed.

It can be seen that a great deal of liaison is required among site personnel. Several persons may be involved with one particular function. For example, the chief engineer, the quantity surveyor, the field engineers, and the section engineers may be involved with the monthly

measurement. This operation alone can require tremendous coopera-
tion and coordination.

The overall engineering functions and duties of a civil engineering
contractor are as follows:

1. Planning
2. Programming
3. Temporary works design
4. Preparing detailed estimates
5. Negotiations
6. Supervising layout of works
7. Supervising work
8. Recording progress
9. Preparing reports
10. Liaison with engineer
11. Quantity surveying
12. Purchasing materials
13. Measurement for monthly certificates
14. Purchasing, operating, and maintaining plant, equipment, and
    transport
15. Work study
16. Quality control
17. Value engineering
18. Cost control
19. Performance measurements and analysis
20. Research and development

General duties and functions are listed as follows:

1. Job procurement
2. Programming and coordinating work of subcontractors
3. Legal work, securities, and bonds
4. Planning and obtaining insurance
5. Ordering, invoicing, and records
6. Measuring and recording costs
7. Management of claims
8. Plant, equipment, and transport records
9. Public relations, labor relations, and union negotiations

10.   Correspondence
11.   Negotiating and operating incentive payment schemes
12.   Coordinating joint ventures when applicable

2.8   THE CONSULTING ENGINEER'S DUTIES AND RESPONSIBILITIES

After the consulting engineer or architect has produced the specifications, the conditions of contract, the bill of quantities, if applicable, the drawings, and a form of tender for the project, and a contractor has been selected to execute the works, the form of contract agreement is signed between the owner and the contractor.

After this event the consulting engineer sometimes only approves temporary works designs, products, and materials, and does not provide a resident staff on site to supervise the work. Other times the consulting engineer, after producing the contract documents and supervising the contract agreement, sets up and provides a resident engineer's staff on site. Alternatively, a regional authority may provide supervisory staff after the consulting engineer has produced the contract documents. Whichever team is resident on site the organization chart is similar to the one shown in Figure 2.4.

The partners in a consulting civil engineer's practice supervise the design and preparation of the contract documents for all the current work that is being carried out. They are involved with job procurement but are bound by certain advertising restrictions. For each particular large project a section engineer is appointed to take charge of feasibility studies, estimating, design, and preparation of the contract documents. Each section engineer heads a design team consisting of qualified and experienced design engineers, technicians, and draftspersons so that the team can produce the drawings required for the contract documents in the time required by the owner. After the contract has been let to a contractor, the office team, although perhaps smaller, continues to function in the production of working drawings such as reinforcement drawings for the various structures.

In the head office a small administrative section takes charge of the clerical, accounting, and personnel duties. This section may also establish a small resource center or library in which technical literature, textbooks, standards, and codes of practice can be loaned to the employees.

If a resident engineer's office is established on site, the resident engineer is the chief person responsible for maintaining the owner's interests on site. The resident engineer ensures that the work is executed according to the conditions of contract, the drawings, and specification. An assistant engineer or deputy engineer may take over part of the resident engineer's job for daily site supervision, the issuing of working drawings, and instructions to the contractor.

Engineering assistants are then allocated to a particular section of the project if the size of the project merits it. These engineering assistants carry out the general daily supervision of their section. This supervision may entail the approval of the layout of a structure, checking that the structure has been built according to the drawings, or planning for the testing of materials. An inspector assists the engineering staff so that each section of a project can be continuously inspected. The inspectors remain on site continuously, inspecting the contractor's work. All their observations are recorded in a diary. This diary, which may be used subsequently for the purposes of claims, should record the weather, the daily progress of each job, and the cause and amount of interruptions and delays to the progress of work. Often preprinted books are used by the inspectors to record their observations. The number, type, and status of each person on each job is recorded as well as the number, type, and size or rating of each item of equipment. Sometimes surveyors instead of engineers are allocated to a section, depending on the amount of laying out.

The duties and responsibilities of the consulting civil engineer's head office staff are as follows:

1. Preparing designs
2. Preparing specifications
3. Preparing conditions of contract
4. Preparing and issuing drawings
5. Supervising bid arrangements
6. Supervising formal agreement of the contract between owner and contractor
7. Ensuring that all necessary instructions have been given
8. Agreement to a master program of work
9. Issuing further instructions and clarifications when necessary
10. Ensuring that all correspondence is promptly dealt with
11. Reporting regularly to the owner
12. Redesigning work when required
13. Ensuring that final drawings contain all amendments for record purposes

14. Recording all necessary information for possible future work or modifications
15. Examining temporary works design
16. Ascertaining the final value of work

If the consulting engineer's organization acts as head of a project management team, it is necessary to coordinate all the work of the various contractors.

The duties and responsibilities of the consulting engineer if the firm continues with on site inspection are as follows:

1. Ensuring that work is executed according to the drawings, specifications, and conditions of contract
2. Recording progress
3. Inspecting materials and workmanship
4. Ensuring that completed work is tested when necessary
5. Recording all measurements and tests
6. Measuring the amount of work done for interim payments
7. Checking the layout of the works
8. Keeping a diary of all the works
9. Arranging visits and performing public relations duties
10. Providing all the facts for claims
11. Continuously reporting back to the head office
12. Ensuring that final drawings show the actual "as constructed" details
13. Issuing further instructions and clarifications when necessary
14. Assessing the contractor's alternative proposals for a particular method

2.9 CONDITIONS OF CONTRACT

The use of the design and construct type of contract has decreased in civil engineering, as owners have preferred the system of competitive bidding. This generally means that a consulting civil engineer is employed to prepare drawings, specifications, and conditions of contract. In some countries there are no standard conditions of contract. Conditions of contract and specifications have been drafted and published in various formats. Steel institutes, highway authorities, concrete societies, and welding societies have produced their own separate specifications. Other types of conditions of contract have been drafted within many firms and altered to suit the local regulations and the current legislation.

There is a need for a standard set of conditions of contract which would cover most general aspects of civil engineering. Additional clauses would be required for specialized items, but a standard conditions of contract would improve communication between the contractor and consulting design engineer. Improved communication is now more essential for projects that are not of the design and construct type.

Apart from the improved communications, the conditions would be less likely to contain mistakes or ambiguities. Engineers responsible for this type of work would soon become familiar with the standard conditions of contract, instead of being unfamiliar with another firm's or authority's set of conditions, at the beginning of a new project. Any slight ambiguities would be resolved once a precedent had been set for all to see. This would have the beneficial effect of permitting the professional engineer to more often resolve disputes instead of resorting to arbitration or employment of a lawyer.

The fifth edition of a document known as "The I.C.E. Conditions of Contract"[5] was issued in 1973 in the United Kingdom. This is the type of document that is necessary where there is not a set of standard conditions of contract existing. The document has been issued and accepted by the Institution of Civil Engineers and representatives from the consulting civil engineering association and the civil engineering contractors. A permanent joint committee continuously reviews the document for revisions and amendments. The document defines, explains, and interprets the essential conditions of contract. Some of the subheadings listed in the document are as follows:

1. Engineer's representative
2. Assignment and subletting
3. Contract documents
4. General obligations
5. Labor
6. Workmanship and materials
7. Commencement time and delays
8. Liquidated damages and limitation of damages for delayed completion
9. Completion certificate
10. Maintenance and defects

2.10 SAFETY[6]

The contractor must take full responsibility for the safety of all site operations and methods of construction except for the design and specifications of the permanent works or the design of any temporary works designed by the consulting engineer. As the project progresses the contractor must have the fullest regard for all persons entitled to be present on site and is obliged to keep the site in an orderly way so that persons on site are not endangered. Warning notices, lights, guards, and barriers should be installed at the contractor's expense when so requested by the consulting engineer or any other competent statutory authority or other authority so that the protection of the works and the safety of site personnel or the general public can be ensured.

Many failures in the past have been due to inadequate technical knowledge. The introduction of new materials or new processes can cause failures or danger unless innovations are thoroughly tested and a proper research program carried out to investigate all the unknown properties. However, as more research programs are carried out, technical knowledge increases and errors become the largest single cause of structural failures, particularly in technologically advanced countries. An error in this context is a gross error or mistake and not a minor calculation error or slight construction inaccuracy. These errors are made by builders, contractors, designers, manufacturers, or the authority responsible for checking design or construction. Usually more than one individual has some responsibility in a gross error. In fact a system of checking should be devised so that it is not solely one individual's responsibility for important civil engineering designs.

In a sense accidents or failures would be more likely today if it were not for the increased technical knowledge among the engineering professions. Factors of safety are being continuously reduced, and modern structural forms have less inherent strength than an arch, for instance, which was used centuries ago. Today most structures are designed by a rational direct method involving a good understanding of mechanics and structural analysis. The purpose of design is to ensure that the structure does not become unfit for use. All relevant limit states, such as ultimate strength, deflection, cracking, and vibration should be considered to ensure that an adequate degree of safety and serviceability is attained. Generally two separate partial safety factors are used for the materials and for the loads.

Good communications and thorough inspection are very important in preventing errors. A lack of competent structural engineering knowledge in the erection stage or design of temporary works has contributed to a disproportionate amount of failures in the past.

Workmanship in some manufacturing processes is particularly prone to failure. Strict control and inspection are required to prevent serious defects in operations such as welding, scaffolding, curing of concrete, particularly during extreme climatic conditions, supports in trench excavation, and gluing laminations in timber construction.

When a structure is complete, a planned system of inspection and maintenance is essential to reduce failure in the long term. A change of use within areas of a multistory building can impose severe overstressing. Machinery, including large computers, can overload a structure. The building is designed for a given loading per unit floor area and this design load should never be exceeded.

Excavations can cause failures in the walls of adjacent buildings. It is therefore essential to assess the foundations and the distribution of foundation loads in adjacent buildings. Inadequate knowledge of the structural integrity of a building that is being demolished can also lead to very sudden structural collapse.

Awareness of safety procedures, trained judgment and technical knowledge of methods and materials, and the experienced professional engineer's consideration of stability can all help to reduce the occurrence of failure and therefore increase safety on site.

CHAPTER 3

PLANNING, PROGRAMMING, AND PROGRESS

3.1 INTRODUCTION

There are many reasons why a system of planning is absolutely essential on a construction site. For example, a very simple job on site involves the construction of a wall or abutment base, as shown in Figure 3.1. The operations, usually referred to as activities in modern programming methods, the trade types, the number of personnel, and the time durations of the activities are shown. It can be seen that some amount of planning is necessary to avoid each gang being in the same area during the construction of the base.

The excavating machine, the timber formwork, the steel reinforcement, the readymixed concrete, and the labor gangs need to be provided on time and planned for well in advance. Frequently there are numerous activities progressing simultaneously and the logical sequence of the activities can be somewhat complicated and obscure. It is then necessary to record a program of work so that interested parties can see beforehand the plan of the intended schedule of events. In the early stages the program can be altered when a feedback of comments is received.

The different parties involved often require different types of programs. The owner of the project, who may not possess an engineering background, often only requires a simple outline program showing the key events in the project. The estimating department of a construction company and the project manager require an overall master program that shows all the main activities of the project. The engineer in charge of a section and the general foreman require a very detailed current program of from 1 to 4 weeks duration so that the daily location of the labor gangs, materials, and equipment can be seen.

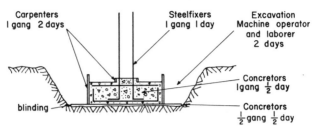

Figure 3.1   Construction of a wall base.

3.2 BAR CHARTS

The use of bar charts for planning in construction is an old and well established method. A bar chart simply shows the period of time in which certain parts of the works are scheduled for construction. A typical bar chart is shown in Figure 3.2. It is simple to understand and is the most commonly used method of programming. A bar chart is nearly always used for the outline program (see Figure 3.2) so that the owner, the contracts manager, the civil engineers, and the foreman can confer in groups or together. The bar chart is an effective means of communication.

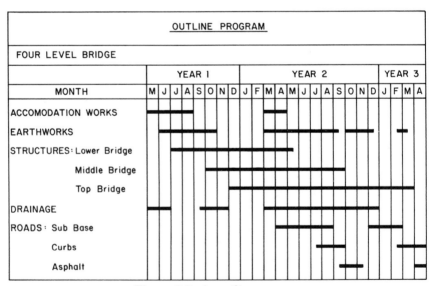

Figure 3.2   An outline program.

Often many activities are presented as one operation; for example, the construction of a long basement slab for a factory may be shown to take 4 weeks as follows:

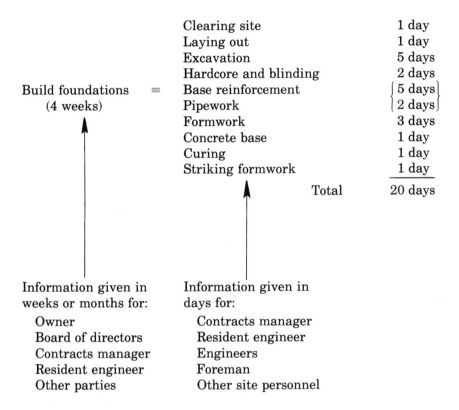

	Clearing site	1 day
	Laying out	1 day
	Excavation	5 days
	Hardcore and blinding	2 days
Build foundations ≡	Base reinforcement	$\begin{cases} 5 \text{ days} \\ 2 \text{ days} \end{cases}$
(4 weeks)	Pipework	
	Formwork	3 days
	Concrete base	1 day
	Curing	1 day
	Striking formwork	1 day
	Total	20 days

Information given in weeks or months for:
  Owner
  Board of directors
  Contracts manager
  Resident engineer
  Other parties

Information given in days for:
  Contracts manager
  Resident engineer
  Engineers
  Foreman
  Other site personnel

Usually, the general information such as "build foundations" is used in the bar chart type of program, the time unit being measured in weeks or months.

It can be seen in the example that two activities overlap and therefore the total time only includes the larger time of the two activities. The other activities are listed in sequence; one activity cannot commence until the previous activity has been completed. The reinforcement, formwork, and pipework must be delivered to site before certain activities can begin. The actual ordering and delivery of these materials is a restraint on the scheduled sequence of activities.

Because of these overlaps, restraints, the linking of activities in sequence, and, in particular, a requirement for determining the critical jobs in the program, a more sophisticated method of planning has been developed, called critical path methods. Sometimes, however, even if the more sophisticated methods are used by the engineers and planners, the outline program is still converted to the bar chart form.

3.3 CRITICAL PATH METHODS

Critical path analysis clearly shows the interdependence and interrelationship of activities in a scheduled order as shown in Figure 3.3. A critical path program may be drawn up by using the following step by step procedure. Reference should be made to the simple example of the construction of a factory concrete floor in Figure 3.4.

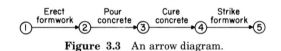

Figure 3.3   An arrow diagram.

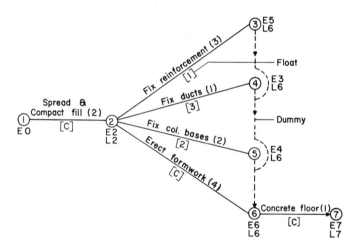

(_) shows estimated time duration of activity

———   dummy maintains sequence and interdependence
        of activities, it is non time consuming

[   ]   total float, the available spare time for
        an activity

E   shows the earliest event time

L   shows the latest event time

[C]   shows the critical path with zero float

Figure 3.4   A simple critical path network.

A critical path analysis is devised as follows:

1. Write down all the activities that are involved. The time durations are not considered yet (and therefore the network cannot be drawn to a time scale at this stage).

2. Decide the logical sequence for the activities by selecting one activity and deciding which activities must precede it, which can be simultaneous, and which must succeed it. Repeat for all activities (Figure 3.5).

3. Estimate and record the time duration of each activity. This estimate can be made from previous experience, from the incentive payments scheme, or from tables.

4. Number the events for easy reference.

5. Remember that an event cannot be reached until all the preceding activities are complete. If there are several activities leading to one event, the earliest that that event can occur is the latest time path. The earliest event time is therefore derived by working forward: $E \rightarrow$.

6. Derive the latest event time by working backward: $\leftarrow L$. Note that continuous reference should be made to Figure 3.4.

7. Deduce the float, the available spare time. Float equals the latest event time minus the earliest event time. Use the larger value of either end.

8. Plot the critical path. This is the path that goes through all the activities with zero float. It is often colored in red to make it prominent.

Two other terms that are frequently used in critical path planning are shown in Figure 3.6. A restraint is a restriction depending on the availability of machines, equipment, materials, or labor. Free float is the amount of spare time available in which a delay will not affect the total float of a following activity.

In practice it is found that certain activities overlap in time. This is a problem in critical path planning, but it can be overcome either by using dummies or by splitting an activity into two parts.

Although the best and most logical critical path program may be drawn up, this may not be the best solution if the resources are required for several jobs at one particular time.

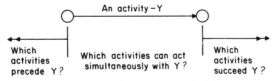

Figure 3.5 Sequence of activities.

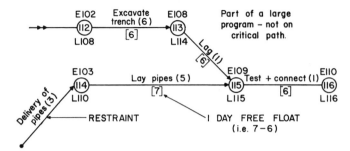

Figure 3.6   Restraint and free float.

3.4   RESOURCE LEVELING

Resource requirements such as labor, materials, machines, and equipment may fluctuate according to the initial program. These requirements may include activities either with float or activities on the critical path. One of the advantages of the critical path method is that the activities that are critical at a certain time are clearly identified. Activities that are not critical, and therefore possess float, are also clearly identified with the amount of float available.

If the total number of persons required for a certain period in a project fluctuates, the noncritical jobs can be rescheduled so that the total number of persons required remains practically constant or level. This reanalysis is called resource leveling. A simplified example is shown in Figure 3.7.

Note that job C, which is on the critical path, is not altered unless absolutely necessary.

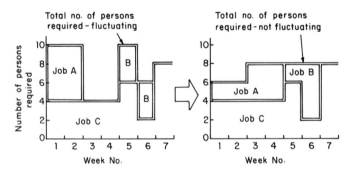

Note: Jobs A and B have float and Job C is on the critical path

Figure 3.7   Resource leveling.

3.5 PRECEDENCE DIAGRAMS

Precedence diagrams or activity on the node diagrams represent a series of interdependent and interrelated activities where each activity is denoted by a block or node instead of an arrow. A precedence diagram block is shown in Figure 3.8.

There are possible advantages of a precedence diagram over the arrow diagram, although the arrow diagram has generally retained its popularity in civil engineering. The precedence diagram method is a simple and straightforward method. Preprinted sheets of blank blocks are generally available, and it is therefore easier, neater, and quicker to draw the diagram. Dummies are not required in the logical sequence and overlaps can be represented easily. A simple comparison is made between an arrow network and a precedence diagram in Figure 3.9.

The simple arrow network shown in Figure 3.4 is reproduced in Figure 3.10 in precedence diagram form.

3.6 NETWORK ANALYSIS BY COMPUTER

In some cases when the number of activities in a project is large, the use of a computer can greatly facilitate the network analysis. The weekly or monthly updating of the program can also be a relatively simple task when a computer package program is used. A more refined

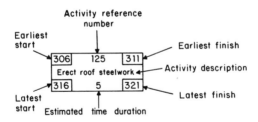

Figure 3.8 Precedence diagram block.

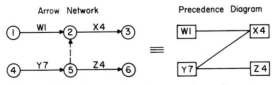

Figure 3.9 Arrow network—precedence diagram.

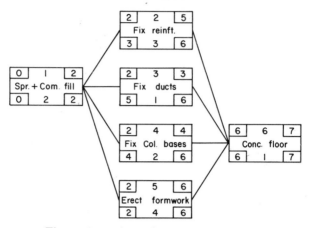

Figure 3.10   A simple precedence diagram.

method of critical path analysis known as PERT (program evaluation and review techniques) is generally associated with network analysis by computer. There are several package programs available and in most of the programs the method of approach is similar.

A package program basically enables the user to easily run a computer program, which is already written, after preparing a network and translating this information onto punched cards. With sophisticated package programs the computer produces a program for the project including the critical path. It can use probability methods for determining the most likely duration of an activity, level and schedule resources, and also update the program at regular periods.

3.7   NETWORK PREPARATION FOR COMPUTER

The following steps should be taken to run a computer package program satisfactorily:

1.   Carefully plan and prepare a program network in the usual way in logical construction sequences
2.   Check the network, preferably by an independent person or group
3.   Copy the network logic onto coding forms exactly as instructed in the user manual
4.   Check the coding forms, preferably by an independent person
5.   Punch the data from the coding forms onto cards

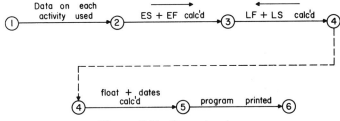

Figure 3.11 Computer steps.

6. Check the data, spaces, column positions, and so on, preferably independently

7. Run the program

By observing many groups attempting to use a computer package program for the first time a logical system has been found to be indispensable. Independent checks, clear instructions, use of the user manual, examples of coding sheets, and examples of the input and output sheets are necessary when users are attempting to run a package program initially.

When the network is prepared for the computer, it is important to remember that the computer follows the same steps as the manual operation. These steps are shown in Figure 3.11.

3.8 INPUT FOR PACKAGE PROGRAM

A simple network is shown in Figure 3.12. The duration of each activity is shown in Table 3.1.

TABLE 3.1 Duration
of Activities

Activity	Duration
A	5
B	10
C	20
D	15
E	5
F	10
G	10
H	10

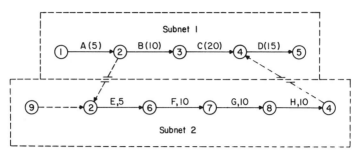

Figure 3.12   A Simple network.

A listing of the input of punched cards for the simple example shown in Figure 3.12 and Table 3.1 is shown in Figure 3.13. The minimum cards required for such a network are as follows:

1 descriptor card [00]

1 dictionary card [01]

1 calendar card [05]

n activity cards [30]

n special event cards [30]

1 EOD card for end of deck

The position of each alphanumeric symbol is significant. The format for the column positions of the data input is specified in the descriptor card, type [00]. An explanation of a typical descriptor card for the Project Management System developed by the IBM corporation is shown in Figure 3.14.

3.9   OUTPUT FROM COMPUTER

By prescribing a definite start date in the data input, the line printer output gives the duration, the float, and the earliest and latest completion date for each activity. A bar chart can also be printed as well as other extra information if stipulated on the input cards.

The application of GERT (graphical evaluation and review technique) and a more detailed description of construction performance control by networks may be found in Ahuja's textbook.[7]

```
//E4022301 JOB (4022,301C,1,1),CHRISTIAN,CLASS=C
// EXEC PMS01
//EDITTIME.SYSIN DD *
00 CHRISTIANTESTM122                4  091419    3138
01  CHRIS1                                                           SUB1
01  CHRIS1                                                           SUB2
05 0080 1 1  1111  21111            422MAY423MAY104JUN               AUHW
10  01MAY80 CHRISTIANTEST1NETWORK                                    CHRIS1
30AS            11          01MAY8001MAY80    BEGIN                  SUB1
30AE     51                                   END                    SUB1
30AI            21                            I.FACETOSUB2ERSUB1      SUB1
30AI            41                            I.FACERSUB2TOSUB1       SUB1
30A AA   11     21     05                     A                      SUB1
30A AA   21     31     10                     B                      SUB1
30A AA   31     41     20                     C                      SUB1
30A AA   41     51     15                     D                      SUB1
30AS            91          08MAY8008MAY80    BEGIN                  SUB2
30AI            21                            T.FACETOSUB2ERSUB1      SUB2
30AI            41                            T.FACERSUB2TOSUB1       SUB2
30A BB   91     21                            DUMMY                  SUB2
30A BB   21     61     05                     E                      SUB2
30A BB   61     71     10                     F                      SUB2
30A BB   71     81     10                     G                      SUB2
30A BB   81     41     10                     H                      SUB2
30A BB   41     10                            DUMMY                  SUB2
30AE     10                                   END                    SUB2
EOD
/*
//PROCESS.SYSIN DD *
01  CHRIS1     0
EOD
]*
//REPORT.SYSIN DD *
*TIMEREO        0102                                                 CHRIS1
/*
//
/*
```

Figure 3.13   Listing of input cards.

3.10   OUTLINE PROGRAM

The outline program is used at the initial stages of the project and is drawn up at the same time as the conceptual plans are drawn. Some of the persons involved in the project at this stage may not be conversant with civil engineering terms and not particularly interested in each individual activity.

Numerous discussions between the owner and the consulting civil engineer take place, and often only the general outline is discussed. For this reason only a relatively small number of operations are used in the outline program. Many activities are contracted into one operation in the same way as described in Section 3.1 for the "build foundations" operation. During the conceptual stage of the project it is likely

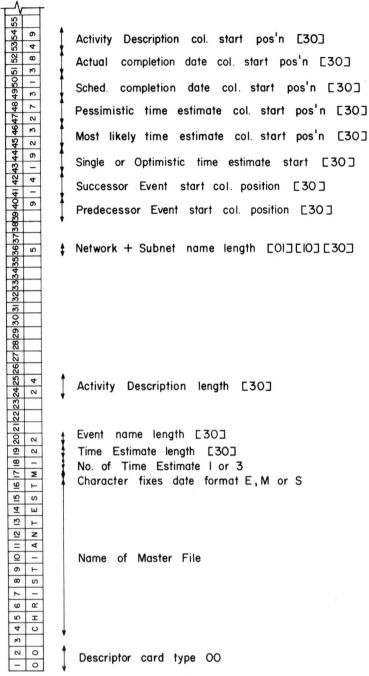

Activity Description col. start pos'n [30]

Actual completion date col. start pos'n [30]

Sched. completion date col. start pos'n [30]

Pessimistic time estimate col. start pos'n [30]

Most likely time estimate col. start pos'n [30]

Single or Optimistic time estimate start [30]

Successor Event start col. position [30]

Predecessor Event start col. position [30]

Network + Subnet name length [01][10][30]

Activity Description length [30]

Event name length [30]
Time Estimate length [30]
No. of Time Estimate 1 or 3
Character fixes date format E, M or S

Name of Master File

Descriptor card type 00

[ ] denotes ' on card type'

Figure 3.14 Explanation of descriptor card.

52

that the original outline program will be progressively altered as the concept develops and more knowledge and information is known about the project.

The final outline program is generally in bar chart form. It enables the start and finish dates, the project duration, and the time of any major constraint to be calculated for use in the tender documents. A typical outline program is shown in Figure 3.2.

3.11 MASTER PROGRAM

The master program breaks down the major operations shown in the outline program into a series of individual activities. On large projects the master program may contain over 1000 activities and is usually represented in network form. The actual dimensions of the master program may vary from 2 to 3 m long by 1 m high. It is therefore impossible to represent this type of program accurately on one page of a textbook. All the activities included in the project are shown on the master program, with each individual activity shown in a little less detail than shown in the detailed program, Figure 3.15.

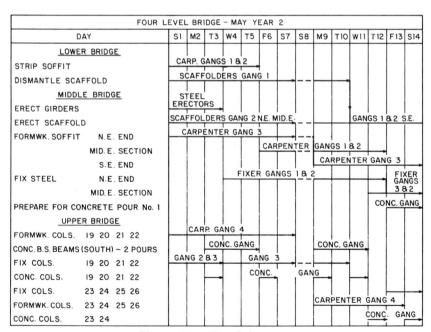

Figure 3.15 Detailed program.

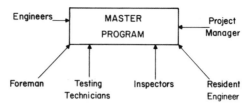

Figure 3.16   Personnel who refer to master program.

The contractor generally draws up the master program, which should be submitted to the consulting engineer as soon as it is complete. Figure 3.16 shows the personnel on site who frequently refer to the master program.

The main contractor can assess from the master program when the following resources are required:

1. Subcontractors
2. Materials
3. Gangs for various trades and other personnel
4. Various types of equipment

The resident engineer is able to determine from the master program when certain supervisory personnel are required and when the working drawings are required.

All programs should take into account the seasonal fluctuations encountered due to wet or cold weather and changes in the number of hours worked each week so that the programs do not become outdated. It is, however, almost inevitable that a program will become out of date. Several factors that frequently cause delays are as follows:

1. Prolonged bad weather
2. Labor troubles
3. Late deliveries
4. Late arrival of subcontractor
5. Insufficient time allowed for deliveries
6. Changes in design and construction

3.12   DETAILED PROGRAM

A detailed program is shown in Figure 3.15. Detailed programs are usually scheduled in 2, 3, or 4 week intervals. Each section of a project has its own separate detailed program, with each program dovetailing into the master program. The detailed program shows the location of each gang and should very accurately schedule all the resources, including labor, materials, and equipment, in a logical sequence that ensures that there are no conflicting resource requirements.

3.13 PROGRESS CHARTS

Programs may be arranged so that the rate of progress can be shown on them. These programs are called progress charts or production diagrams. The rate of progress may be recorded by means of a second line on the program which records the progress data, a string line, or by a method similar to that shown in Figure 3.17.

If data are recorded on the program, the unit of measurement may be in number, length, area, or volume depending on what is appropriate.

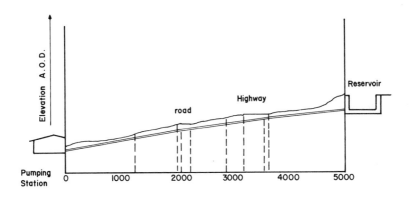

2 ≡ end of month 2

Figure 3.17 A progress chart.

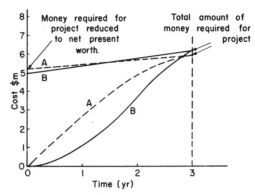

Figure 3.18   Real cost versus time.

3.14   RELATIONSHIP BETWEEN PROGRAM AND REAL COST OF PROJECT

Programs are important to determine the real cost of a project. To determine the real cost of a project the total amount of money required for the project should be reduced to the net present worth at the beginning of the project, as shown in Figure 3.18. The graph shows the running costs of the same project scheduled by two different programs A and B. It can be seen that the line that reduces the total amount of money required for the contract to the net present value is steeper for program B than for program A. The reason for this is that less money is used in the first 2 years for program B. The program therefore has a direct effect on the real cost of a project. The fundamental principles of net present worth and discounted cash flows are further explained in Chapter 4.

3.15   TYPES OF LIMITED RESOURCES

To realize a profit the contractor must level and limit resources. For example, materials for temporary works should be used as many times as possible and not just scheduled for one particular activity. Therefore for each day on a construction site the labor, machines, equipment, and materials are limited.

Limited resources should be allocated to activities, in order of priority and preference that:

1. Are on the critical path
2. In progress
3. Contain the least float
4. Require the largest number of resources

Example 3.7 contains an exercise on resource limitations.

3.16 WORKED EXAMPLES

Example 3.1

Find the earliest and latest event times, the amount of float on each activity, and the critical path of the following example, which describes the construction of a wall from the concreting of a base to the concreting of the stem:

Activity	No. of days
Pour concrete in base	1
Fix steel in stem	3
Fix pipes in stem	1
Strike base formwork	3
Backfill around base	2
Erect stem formwork	3
Pour concrete in stem	1

Answer 3.1

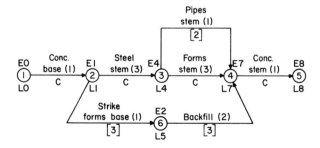

Example 3.2

Represent the following arrow diagram by a precedence diagram.

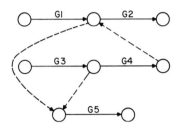

Answer 3.2

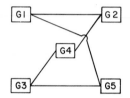

Example 3.3

An activity *JA* occupies day 51 to day 55, another activity *JB* occupies day 53 to day 58; that is, there are 3 days of overlap. Represent the above by:

(a)   An arrow diagram
(b)   A precedence diagram

Answer 3.3
(a)

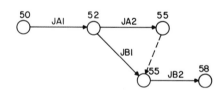

(b)

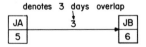

Example 3.4

The estimated times, in days, of all the activities for the construction of a prefabricated temporary building are as follows:

Water supply
Lay out trench	2
Deliver pipes	3
Excavate trenches	6
Lay pipes	5
Test and connect to mains	1

Soil pipe network
Lay out trench and manholes	2
Deliver pipes	2
Excavate trenches	8
Excavate for drop manholes	3
Construct drop manholes	10
Lay pipes	7
Test and connect to mains	1

Electricity connection, total time	4

Prefabricated building
Construct foundations	2
Deliver prefabricated units	8
Erect prefabricated units	1
Complete interior of building and so on	6

Allow a minimum time lag of 1 day after excavation to finish laying the pipes.

It may be assumed that the water supply and soil pipe operations are independent of each other.

It is possible for all types of excavation to proceed simultaneously.

Show the critical path diagram including total float times, and thus determine the earliest event time for completion of the project.

If the delivery of the prefabricated units is delayed by 3 days, what effect will this have on the time for completion of the project?

Answer 3.4

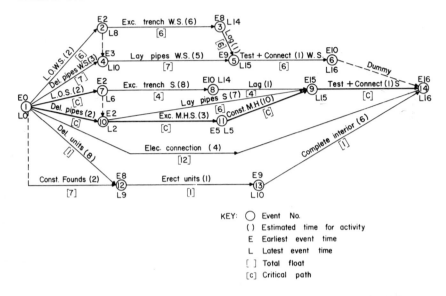

KEY: ◯ Event No.
( ) Estimated time for activity
E Earliest event time
L Latest event time
[ ] Total float
[c] Critical path

The earliest time for completion of the project is 16 days. If the delivery of the prefabricated units is delayed by 3 days, the critical path jumps to (1) → (12) → (13) → (14). The earliest time for completion is then 18 days, that is, original time + delay − float (16 + 3 − 1) = 18. Note that this answer assumes that the trenches can be backfilled after occupation of the building and also that the connection of the pipes to the building is independent of the construction of the building.

Example 3.5

Repeat Example 3.4 using the precedence diagram method.

Answer 3.5

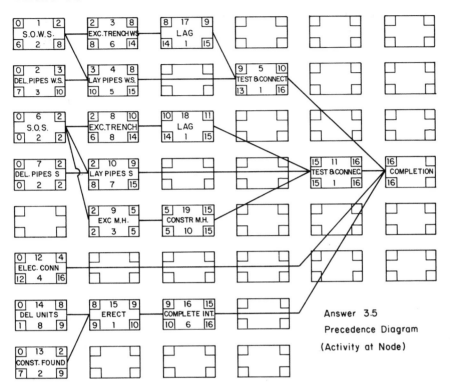

Answer 3.5

Precedence Diagram

(Activity at Node)

Example 3.6

The estimated times in days of all the activities in connection with the construction of a gas filling station are as follows:

Clear site	3
Laying out	1
Office	
Excavation	3
Foundations	2
Windows	2
Walls	7
Roof	3
Carpentry	4
Plumbing	4
Electrical work	3
Plaster boards	4
Forecourt	
Excavation	3
Concrete slab	1
Brickwork	3
Tank installation	3
Complete forecourt base	3
Fix pumps	1
Install canopy and furnish forecourt	7
Delivery of tanks	7
Underground services	
Underground services installation	8
Lay hardcore fill	4
Underground equipment for pumps installation	2
Painting of forecourt and office	8
Clear site	4

It should be assumed that the windows can be installed while the office walls are being built. The carpentry, plumbing, and electrical work can proceed simultaneously after the office roof has been built. The underground services work should also be assumed to be proceeding at the same time as the forecourt work.

Show the critical path diagram, including float times, and thus determine the earliest time for completion of the project.

Answer 3.6

Earliest Time – 14
Latest Time – 16
(Float) – (2)

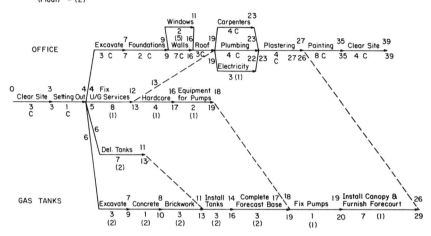

Example 3.7

This is an example on resource limitations. Two abutments are to be constructed in three stages: the base, the lower stem, and the upper stem. All the activities for abutment No. 1 which should be considered are as follows:

Activity	Duration (days)	Code
Formwork to base	4	FB
Rebar base	2	RB
Concrete and cure	2	CCB
Rebar lower stem	3	RLS
Formwork to lower stem	4	FLS
Concrete and cure	3	CLS
Rebar upper stem	3	RUS
Formwork to upper stem	4	FUS
Concrete and cure	3	CUS

The same times should be assumed for abutment No. 2 as for abutment No. 1. One extra carpenter and two laborers are available on the site for striking the forms, so this activity can be ignored. One set of forms for the use of both bases should be assumed unless otherwise stated. There are no limitations on the rebar gangs.

Draw an arrow network for the following cases, which prescribe different resource limitations:

CASE 1   Use the same forms on the lower and upper stems of each abutment. Complete abutment No. 1 first and as soon as possible. More than one carpenter gang is available.

CASE 2   Use the same forms on the lower and upper stems of each abutment. Complete abutments 1 and 2 as soon as possible. More than one carpenter gang is available.

CASE 3   Use one carpenter gang only with no limitations on the provision of formwork. Plot a histogram of the carpenter gang working days. Complete both abutments as soon as possible. Note that two sets of base forms may be used in this case.

CASE 4   Use one carpenter gang and one set of forms only for the abutment stems 1 and 2, lower and upper. Complete both abutments as soon as possible.

CASE 5   Assume that there are no limitations on the number of carpenter gangs and the sets of formwork, although no more than one gang is allowed in one particular area. Complete both abutments as soon as possible.

Answer 3.7

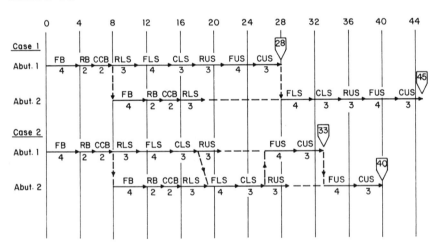

CASE 1   Abutment 1 completed in 28 days.
Abutment 2 completed in 45 days.

CASE 2   Abutment 1 completed in 33 days.
Abutment 2 completed in 40 days.

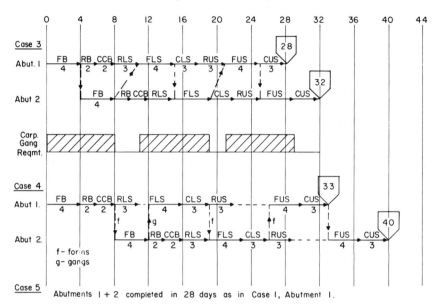

CASE 3   Abutment 1 completed in 28 days.
Abutment 2 completed in 32 days.

CASE 4   Abutment 1 completed in 33 days.
Abutment 2 completed in 40 days.

CASE 5   Abutment 1 completed in 28 days.
Abutment 2 completed in 28 days.

Example 3.8

This is an example on resource leveling and scheduling. Level the resources and schedule the following jobs.

		No. of working days	Earliest start	Initial programmed duration	Float
	Job A	28	0	4	6
	Job B	20	2	2	7
Critical path	Job C1	20	0	5	C
	Job C2	20	10	5	C
	Job C3	20	20	5	C
	Job D	56	9	14	2
	Job E	30	15	3	4

All jobs concern the same trade and four person gangs should be assumed.

Answer 3.8

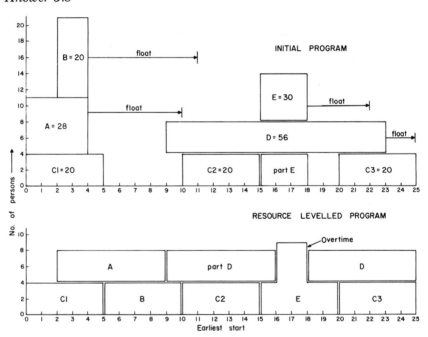

The resource leveled schedule is shown in the histogram. Often, in practice, the solution is not so neat.

CHAPTER 4

PROJECT APPRAISAL

4.1 PRESENT VALUE

If a sum of money is borrowed from a bank or elsewhere for the construction of a project, the loan company charges interest at a given rate each year. For instance, if $100 is borrowed at an interest rate of 10%, the amount owing at the end of one year is $110, which includes $10 interest. At the end of the second year the amount owing is calculated on the amount originally borrowed plus the interest. Hence the amount owing at the end of the second year will be $110 × 1.10 = $121, which includes $21 compound interest.

The amount owing therefore accumulates each year as follows:

Beginning of project, amount of loan	$100.00
Interest after 1 year (100 × 10%)	10.00
Amount after 1 year	110.00
Interest after 2 years (110 × 10%)	11.00
Amount after 2 years	121.00
Interest after 3 years (121 × 10%)	12.10
Amount after 3 years	133.10
Interest after 4 years (133.10 × 10%)	13.31
Amount after 4 years	$146.41

The following mathematical formula can be used in calculations involving compound interest:

$$S = A(1 + r)^n \qquad (4.1)$$

where $S =$ the compound total

       $A =$ the original amount, principal, or present value

       $r =$ the interest rate per annum

       $n =$ the number of years

For example, if \$100 is borrowed from a bank at an interest rate of 10%, the amount owed after 4 years is:

$$
\begin{aligned}
S &= A(1 + r)^n \\
&= 100(1 + 0.1)^4 \\
&= 100 \times 1.4641 \\
&= \$146.41
\end{aligned}
$$

This amount may be calculated using the formula but is often read from prepared tables that give the terminal value of a single sum at compound interest rates. As the interest rate is likely to fluctuate frequently, the tables usually give values in 0.5% interest rate increments from 1.0% up to about 20%.

The inverse of compounding is known as discounting. If the same symbols are used, the following mathematical formula can be used in discounting calculations:

$$
A = \frac{S}{(1 + r)^n} \tag{4.2}
$$

Again, as an alternative to using the above formula there are tables available which give the present value of a single sum.

Example 4.1

\$146.41 is required for a project in 4 years' time. What is the amount of money that has to be invested at present to accumulate this amount if the interest rate is 10%?

Answer 4.1

$$
\begin{aligned}
A &= \frac{S}{(1 + r)^n} \\[2mm]
&= \frac{146.41}{(1 + 0.1)^4} \\[2mm]
&= \frac{146.41}{1.4641} \\[2mm]
&= \$100
\end{aligned}
$$

The $100 is known as the present value or present worth.

The preceding example demonstrates in a very simple form the more sophisticated techniques used in modern investment appraisal which, because of high interest rates, takes into account the significance of the interest rate and therefore the time value of money. It is therefore usual in making economic comparisons to reduce all the expenditures to the present value.

Example 4.2

A project will take 5 years to complete and there will be no interim return on investment. If contractor X submits a bid of $12.5 million and contractor Y submits a bid of $12.0 million, determine which bid will be the cheaper to the owner, using net present value calculations, if the amount of work carried out and paid for each year is as follows:

	Contractor X	Contractor Y
End of 1st year	1.000	3.000
End of 2nd year	2.000	4.000
End of 3rd year	2.500	2.000
End of 4th year	3.000	2.000
End of 5th year	4.000	1.000
	12.500	12.000

In $ million.

Assume a discount rate of 10%.

Answer 4.2

The present value calculations for the amount of work each year is as follows:

	Discount factor	Contractor X	Contractor Y
End of 1st year	0.90909	0.909	2.727
End of 2nd year	0.82645	1.653	3.306
End of 3rd year	0.75131	1.878	1.503
End of 4th year	0.68301	2.049	1.366
End of 5th year	0.62092	2.484	0.621
Present value		8.973	9.523

All amounts are in $ million, to the nearest $1000. The discount factor is $1/[(1 + r)^n]$.

The preceding example is unrealistic because the rate of progress is usually slower at the beginning and end of a contract. The rate of progress and amount spent monthly generally follow an S curve pattern. This is demonstrated in Section 4.3. However, the example dramatically illustrates that the lower bid price submitted by contractor Y is not the cheapest contract to the owner when net present values are considered.

As the timing of the execution for each part of the civil engineering works is important, it is essential to have an outline program when estimating the present value of a contract.

4.2 DISCOUNTED CASH FLOW (DCF)

If the net present value at the end of a project is positive, the project will be profitable and increase the resources of a company.

Example 4.3

A particular project will last for 4 years. Determine whether the project should be accepted if the initial expenditure is $1.0 million and income from the project will be:

After 1st year: 0.200

After 2nd year: 0.300

After 3rd year: 0.500

After 4th year: 0.400

Assume a discount rate of 8%.

Answer 4.3

Year end	Expenditure	Income	Discount factor	Discounted cash flow −	+	Net
0	1.000		1.00000	1.000		−1.000
1		0.200	0.92593		0.185	−0.815
2		0.300	0.85734		0.257	−0.558
3		0.500	0.79383		0.397	−0.161
4		0.400	0.73503		0.294	+0.133

All amounts in $ million to the nearest $1000.

The resources will increase by $133,000 approximately, and therefore the company considering the project would probably proceed with it unless other equally important projects under review were considered to be more profitable.

This method of finding the net present value by using discounting methods is often referred to as a discounted cash flow method.

DCF methods may be used to compare the cost of two machines made by different manufacturers which would be able to perform similar operations on site.

Example 4.4

Compare the cost of operating a machine *A* which will be written off in 6 years with a machine *B* which will be written off in 5 years. Assume a discount rate of 7%. To simplify the calculations taxes, insurances, overheads, and operator's costs may be assumed to be the same for each machine.

Machine	Initial cost ($)	Maintenance cost p.a. at year end ($)	Life (years)
A	100,000	5000	6
B	80,000	6000	5

Answer 4.4

		Machine *A*		Machine *B*	
Year	Discount factor	Expenditure ($)	DCF ($)	Expenditure ($)	DCF ($)
0	1.000	100,000	100,000	80,000	80,000
1	0.935	5,000	4,673	6,000	5,607
2	0.873	5,000	4,367	6,000	5,241
3	0.816	5,000	4,081	6,000	4,898
4	0.763	5,000	3,814	6,000	4,577
5	0.713	5,000	3,565	6,000	4,278
6	0.666	5,000	3,333		
			123,833		104,601

Therefore assuming that taxes, fuel, operator wages, and so forth are the same in each case and are not considered in the comparison, the present value cost of machine *A* is $123,833/6 = $20,639 p.a. The present value cost of machine *B* is $104,601/5 = $20,920 p.a.

In calculating the present value of a regular amount each year at a fixed interest rate, it is unnecessary to calculate the DCF each year. For example, the present value A of a fixed sum F each year over n years at a fixed annual interest rate $r\%$ is as follows:

At the end of 1 year $\quad A_1 = \dfrac{F}{1+r}$

At the end of 2 years $\quad A_2 = \dfrac{F}{1+r} + \dfrac{F}{(1+r)^2}$ $\qquad\qquad$ (4.3)

At the end of n years $\quad A_n = F\left[\dfrac{1}{1+r} + \dfrac{1}{(1+r)^2} + \cdots + \dfrac{1}{(1+r)^n}\right]$

A_n is therefore a geometrical progression in which there are n terms. If equation 4.3 is multiplied by $(1+r)$ we have

$$A_n(1+r) = F\left[1 + \dfrac{1}{1+r} + \cdots + \dfrac{1}{(1+r)^{n-1}}\right] \qquad (4.4)$$

and by subtracting equation 4.3 from equation 4.4

$$A_n = \dfrac{F}{r}\left[1 - \dfrac{1}{(1+r)^n}\right] \qquad (4.5)$$

Thus in the preceding example the net present value of the $5000 maintenance cost of machine A at the end of 6 years at an interest rate of 7% can be calculated using equation 4.5 as follows:

$$
\begin{aligned}
A_n &= \dfrac{F[1 - 1/(1+r)^n]}{r} \\[2mm]
&= \dfrac{5000 \cdot [1 - 1/(1.07)^6]}{0.07} \\[2mm]
&= \dfrac{5000 \cdot (1 - 0.66)}{0.07} \\[2mm]
&= 5000 \cdot 4.7666 \\
&= 23{,}833
\end{aligned}
$$

This latter calculation is simpler, particularly as tables are available which give the numerical value of the $\{1 - [1/(1+r)^n]\}/r$ factor.

Generally the tables are entitled "present value of a regular series" or "present value of an annuity."

4.3 PROJECT COSTS

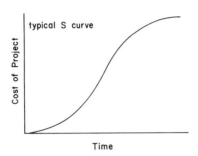

Figure 4.1   Cost of project versus time.

The graph of the cumulative cost of a civil engineering project when plotted against time usually describes an S curve, as shown in Figure 4.1.

The curve is typical of a project, where the rate of progress is slow at the beginning and slows down again at the end of the job but where the rate of progress is fast in the middle part of the job. In some countries the rate of progress depends on the seasonal variations. In some cases the project may come to a complete closedown during periods of inclement weather. Obviously in these cases the cost against time graph is not represented by an S curve.

Example 4.5

An owner requires a project to be completed in 4 years. The project will take 1 year to complete, and a discount rate of 10% should be assumed. Assuming that payment will be made when the job is complete and that there will be no return on investment until after the fourth year, show graphically the present value if the job is completed in:

(a)   The first year at a cost of $8 million
(b)   The second year at a cost of $8.5 million
(c)   The third year at a cost of $9 million
(d)   The fourth year at a cost of $9.5 million

Answer 4.5

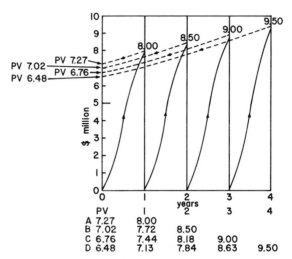

Figure 4.2 Present value of projects undertaken in different years.

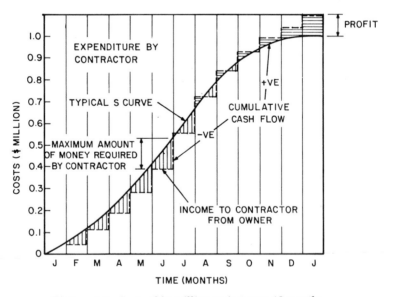

Figure 4.3 Costs of $1 million project over 12 months.

A cost against time of a $1.1 million project over 12 months is shown in Figure 4.3. The S curve shows the expenditure by the contractor and the stepped lines, shown dashed, show the income to the contractor (paid by the owner) in the form of interim monthly payments. If there is no return on investment until after the project is complete, it is obvious that the owner will require $1.1 million to finance the project. However, the contractor will only require finance for the amount of money represented by the maximum height between the S curve and the dashed stepped lines shown in Figure 4.3.

4.4 COMPETITIVE BIDDING

Quite often package deal contracts or negotiated contracts are used, but the most common type of contract in the civil engineering industry is the type where the consulting engineer, on behalf of the owner, prepares tender documents in which several contractors competitively bid for the project.

The competitively bid contract results in a fair competition for the project and, providing it is felt that the particular company is capable of completing the project, the contract is usually awarded to the company that submits the lowest tender. However, if the duration of the contract is to be several years, the consulting engineer should satisfy himself that the time value of money has been considered for the program that the contractor intends to follow.

There are however several important disadvantages in the method of competitive bidding. If, for an example, eight contractors submit a priced bill of quantities when they return the tender to the engineer, only one can be successful. This means that a considerable amount of time and money spent by the seven unsuccessful estimating departments has been virtually wasted.

If a graph is plotted for the value of the bid price against the number of tenders, the graph approximates to a normal distribution curve. Several bids are near the average value and these are likely to be most accurate. Now the successful bid price is usually the lowest, resulting in a reduced profit for the company and may lead to the necessity of a great amount of work put into dubious claims to make a fair profit from the project.

However, in spite of these disadvantages the competitively bid contract works reasonably well and is generally the best available at the present time.

4.5   OTHER FACTORS AFFECTING CASH FLOWS

Although taxation can have a considerable effect on the cash flow of a construction company, the assessment of tax liability has been ignored to facilitate the explanation of discounted cash flow calculations. Tax assessment is difficult for two reasons. First, each country has its own individual methods of taxation, and these methods can be very different when comparing one country with another and attempting to explain a general method. Second, even within one country or state the amount of taxation is likely to vary annually because it depends on the most recent budgetary measures.

In times of inflation the inflation rate obviously affects cash flows.[8] Inflation may be accounted for by substituting an apparent interest rate r_a for the interest rate r in the following equation:

$$(1 + r_a) = (1 + r)(1 + r_i) \qquad (4.6)$$

where r_i is the rate of inflation.

In periods of very high inflation, however, it is increasingly difficult to run a contract on a fixed price basis. At the tender stage all the estimates may contain an extremely large contingency to counteract the effect of inflation. To prevent this, index-linked variation of price formula are used. Certain important cost elements are isolated and are weighted according to their overall proportion in the cost makeup. A published index is then used to vary each cost element over the contract period. The contractor can therefore recover costs that are higher than the original estimate, the yield is therefore more predictable, and the risk in forecasting is reduced. The indices are published each month and usually remain unchanged for 3 months when confirmed final figures are published.

Example 4.6

If the price index for the cost of providing and maintaining constructional plant and equipment was 260.0 at the time of the tender, and during a certain valuation period the plant element cost was £60,000 and the price index was 270.0, determine the amount to be paid to the contractor as a result of price fluctuation if 15% is considered to be a nonadjustable element.

Answer 4.6

Amount of price fluctuation to be paid:

$$0.85 \left(60,000 \times \frac{270.0 - 260.0}{260.0} \right) = £1961.54$$

4.6   TIME VALUE OF MONEY

Engineers in construction form one of the largest spending professional groups. The manner of spending and the allocation of money require difficult and complex decisions. The construction industry is an extremely important sector of a country's economy in terms of employment, output, and consumer. It can account for more than 50% of capital investment and utilize up to about 10% of a nation's labor force. It is therefore inappropriate for the ultimate decision on the allocation and timing of funds to be a political decision rather than an engineering decision based on sound engineering practices and planning principles.

The cost factors of a project include the following:

1.   Labor
2.   Equipment
3.   Materials
4.   Indirect costs
5.   Time

Cash flow is an important part of the time factor. The cash flow has an important effect on the return on capital, profits, and future maintenance programs. The net present value and discounted cash flow analyses, described in Sections 4.1 and 4.2, are important methods in time-cost evaluations.

Figure 4.4 shows a graph of the overall total cost of construction against time. The overall cost is the sum of the indirect costs and the direct costs. The direct costs consist of the labor, equipment, and materials cost factors.

Indirect costs can include the following:

1.   Engineers' salaries
2.   Surveyors' salaries
3.   Office staff wages
4.   Wages of nonproductive personnel
5.   Cost of site accommodations

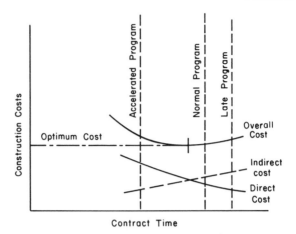

Figure 4.4   Time-cost optimization.

6.   Cost of other facilities
7.   Insurance
8.   Cost of services

It should be noted that the material cost on a construction site is generally not dependent on time, apart from inflationary effects, and could therefore be represented by a horizontal line on the graph. The graph shows that indirect costs increase with time whereas direct costs tend to decrease with time, partly because of the reduction in overtime and incentive payments, and also partly because of the more efficient planning and programming of a lesser number of personnel and machines. The optimal cost often is achieved in a time period less than the normal rate of working. With computer package programs such as PERT now readily available, see Section 3.6, it is easier to predict the theoretical optimal cost. This prediction, however, in spite of the many refinements in the package program, is based on many imponderable factors in the construction industry, such as the weather. The type of graph shown in Figure 4.4 is frequently referred to as a regression curve. It shows that construction costs increase if the program is accelerated and also if the program lags behind schedule.

There are several stages in the planning and construction of a civil engineering project which account for the overall time and for which the promoter must spend money. These component parts of a project are shown in Figure 4.5. The maintenance period is omitted from the schedule.

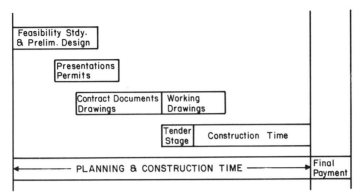

Figure 4.5 Time and cost components of a project.

The profitability of a project for the promoter depends on its overall cost, the probable life of the works, and the annual maintenance cost. For heavy civil engineering works such as dams and tunnels and masonry structures the life expectancy is 100 years or more, the average annual maintenance cost being about 0.1 to 0.2% of the capital cost. Most bridges and buildings are expected to last for about 40 to 60 years and the average annual maintenance cost varies between about 0.2 and 1.0% of the capital cost.[8] Steel bridges tend to have higher maintenance costs than reinforced or prestressed concrete bridges. Stationary machinery has a fairly short life of about 25 to 35 years with a high average annual maintenance cost of between 2.0 and 5.0% of the capital cost. The depreciation of mobile plant on a civil engineering site is usually calculated on a 5 or 6 year life and the annual operating cost, including fuel, can be in the region of 30 to 40% of the purchase price.

Further worked examples on the time-cost aspects in civil engineering can be found in Section 4.13.

4.7 LOAN REPAYMENT

The repayment of loans to the lending organization is often referred to as redemption. The manner in which a loan may be repaid over a specified number of years can vary but there are two common methods. One method permits the repayment of equal installments of the capital and all the annual interest that has accrued on the outstanding balance for that particular year. This method is often referred to as straight line redemption. The second method specifies that the loan should be repaid in equal annual increments covering both the capital and interest. In this method it is assumed that a loan of $\$A$ initially is reduced by $\$X$, which is a proportion of both the amount borrowed and the total interest over n years, the period of the loan. The outstanding debt at the end of each year can be calculated as follows. The outstanding debt at the end of

$$
\begin{aligned}
\text{year } 1 &= A(1+r) - X \\
\text{year } 2 &= [A(1+r) - X](1+r) - X \\
&= A(1+r)^2 - X(1+r) - X \\
\text{year } 3 &= [A(1+r)^2 - X(1+r) - X](1+r) - X \qquad (4.7) \\
&= A(1+r)^3 - X[(1+r)^2 + (1+r) + 1] \\
\text{year } n &= A(1+r)^n - X[(1+r)^{n-1} + (1+r)^{n-2} \\
&\quad + \cdots + (1+r) + 1]
\end{aligned}
$$

As the outstanding debt is paid off after n years expression 4.7 is zero. Therefore

$$
A(1+r)^n = X[(1+r)^{n-1} + (1+r)^{n-2} + \cdots + (1+r) + 1]
$$

Hence

$$
A = X[(1+r)^{-1} + (1+r)^{-2} + \cdots + (1+r)^{-n+1} + (1+r)^{-n}] \quad (4.8)
$$

By multiplying expression 4.8 by $(1+r)$ we have

$$
A(1+r) = X[1 + (1+r)^{-1} + \cdots + (1+r)^{-n+2} + (1+r)^{-n+1}] \quad (4.9)
$$

and subtracting expression 4.8 from expression 4.9 we obtain

$$
Ar = X[1 - (1+r)^{-n}] \qquad (4.10)
$$

Therefore

$$
X = \frac{Ar}{\left[1 - \dfrac{1}{(1+r)^n}\right]} \qquad (4.11)
$$

4.8 NETWORKS AND CASH FLOW FORECASTING

The cash flow of a company is computed from the sum of the net cash flows from all the projects minus indirect overheads such as head office costs, payments to shareholders, and investment. The cash flow forecast for a project is generally extremely difficult due to the many inherent imponderable factors on a construction site.

The lower part of Figure 4.6 shows a graph of cash against time based on the program shown above the graph. For reasons of clarity an

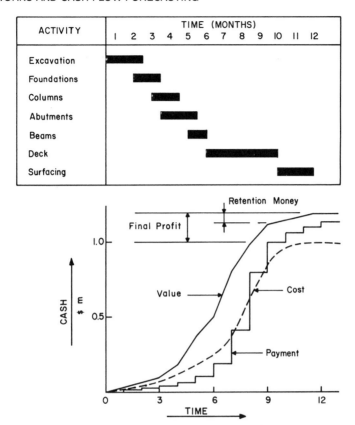

Figure 4.6 Networks and cash flow forecasting.

outline program is shown, but generally a master program is used in cash flow forecasting to ensure a greater degree of accuracy. Three lines are plotted on the graph. The estimated value, which is the payment the contractor eventually receives from the owner, plotted against time, may be obtained from the priced values in the bill of quantities in conjunction with the program. Obviously if the project is the type of contract that does not possess a bill of quantities, the process is much more difficult and probably less accurate.

Analysis of the program should indicate the cash flow lag between the value of the work completed by the contractor and the actual interim payments that he will receive. This analysis enables the stepped payment line to be plotted on the graph. The engineer retains or holds back a small percentage of the money to ensure that the project is satisfactorily completed and maintained according to the conditions of contract. This retention or hold back money should be considered in any cash flow forecast. The contractor spends money for labor, materials, plant, and indirect costs as the project progresses, and these payments, again by referring to the program, can be predicted and plotted on the graph.

Over recent years attempts at cash flow forecasting have become more common as companies have realized the importance of cash flows and the time value of money. These forecasts are often extremely difficult and time consuming. Sometimes accurate forecasts are impossible because of the nature of the project. For these reasons some companies overcome the difficulty by using S curves of value against time based on previous experience and previous case histories. Project values and inflation are dealt with on a pro rata basis. This enables an extremely rapid, though less accurate, cash flow forecast to be made. If the predicted percentage profit or markup is deducted from the value versus time curve, the cost versus time curve can be plotted. This curve generally displays characteristics similar to those obtained from much more rigorous analyses. However, very large and lengthy construction projects often have no meaningful precedent, in which case rigorous analyses with the aid of computer package programs are essential if accurate cash flow forecasts are required.

The payment typically lags behind the cost at the beginning of the project for various reasons such as the time for payment to reach the contractor after measurement and agreement. This lag determines the amount of money that is required to be borrowed for the completion of the project so that the contractor does not encounter any interim cash flow problems. To alleviate such cash flow problems the contractor, during the tender stage, might decide not to have the same percentage profit on all items but to weight items that will be completed at an early stage in the contract with an above average percentage profit.

4.9   RISK ANALYSIS

Risk analysis is essentially a method of determining the estimated costs against time in a contract and assessing the chances of the costs being above or below the estimated values. The opinion of a civil engineer experienced in the type of construction project involved is essential. The number, size, and type of activities play an important role in the variations predicted between the maximum, estimated, and minimum costs. The margin of the probable maximum cost above the

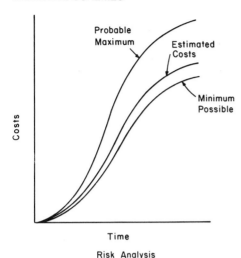

Risk Analysis   **Figure 4.7** Risk analysis.

estimated cost is likely to differ from the margin of the minimum possible cost below the estimated cost. These margins are often extremely difficult to assess, but the use of a computer package program can make the job less onerous even if greater accuracy is not achieved.

Figure 4.7 shows a graph of costs against time. It is worth noting that the minimum possible costs may have a definite value but, with all the imponderable factors prevailing on a construction project, the probable maximum cost cannot be accurately defined. For instance, the final costs of some civil engineering projects have been more than 100% greater than the estimate.

4.10 INCENTIVE SCHEMES

Many types of incentive schemes are used in the construction industry. Some employees find more attractive conditions of working, compared with slightly higher salaries, sufficient inducement to remain with a company. Other unmeasured incentives offer money in addition to basic salary or basic rates of pay without measuring the actual amount of work done. Extra payments may be made in the form of "plus rates," standing bonus, or other benefits such as higher overtime rates or extra pension and insurance rights. A partially measured or assessed incentive scheme may make extra payments, individually or collectively, based on profit or increased output.

The most common type of incentive payment or bonus scheme arranged with operatives on a construction site is the direct payment type of scheme. Payment is made when the basic targets, that is, the normal rates of output per hour, have been exceeded. There are many variations to the scheme. The output rates per hour are obtained from work study evaluations and from previous personal experience. The rates are usually agreed to and fixed by the assistant project manager or engineer and the foreman. In the United Kingdom many operatives work on an incentive scheme. In the United States and Canada, where basic rates of pay are relatively higher, incentive payment schemes are less common. Rates of pay in the construction industry compared with other industries are also higher in North America.

The basic target should be specified for general items of work and should also be simply explained. Incentive payments are usually calculated separately for each gang. All the work done is measured and calculated on an hourly basis, but the bonus is calculated in weekly periods. Schemes can vary considerably, but one type of scheme is so arranged that operatives are paid for any hours that are saved during the week at 50% of the average basic rate of pay for the gang. The basic rate of pay for the gang is generally used in the calculations irrespective of any extra payments that may be made before the bonus is assessed. Distribution to individual operatives is paid on a share system, the shares depending on the number of hours worked in the week and the basic wages earned by each operative. The chargehand's time is considered in the scheme, and payments are only made when the work is satisfactory, with time taken to make good defective work included in the bonus calculations.

Rates of output for a steel fixer's bonus scheme are shown in Table 4.1. It is important to note that there can be many variations in the numerical values.

Obviously the rates of output differ from one job to another. The shape and the quantity of the reinforcement affect the output rate for the cutting and bending, and the location affects the ease of fixing a rebar. The targets shown in Table 4.1 include all labor and transportation of the rebars and apply to all types of reinforcement.

Rates of output for a carpenter's bonus scheme are shown in Table 4.2. Again, it should be emphasized that the numerical values can vary according to site conditions and the type of bonus scheme. The fixing of the formwork includes all work associated with the adjusting to line and elevation as well as the propping, strutting, cleaning, repairing, and oiling of the formwork. The area of formwork considered is the area in contact with the concrete.

TABLE 4.1   A Steel Fixer's Bonus Scheme

| Bar diameter (mm) | Basic target (kg/man-hr) | | |
	Fix	Cut	Bend
10	15	75	25
12	25	110	35
16	35	150	50
20	40	225	75
25	50	300	100
32, 40, and 50	60	400	125

TABLE 4.2   Carpenter's Bonus Scheme

Item	Description	Basic target/ man-hr
1.	Make formwork to beam sides not exceeding 1 m in depth (high grade finish)	0.9 m²
2.	Make formwork to beam sides not exceeding 1 m in depth (ordinary finish)	1.2 m²
3.	Make formwork to walls and beam sides exceeding 1 m in depth (high grade finish)	1.1 m²
4.	Make formwork to walls and beam sides exceeding 1 m in depth (ordinary finish)	1.4 m²
5.	Fix formwork to walls and abutments	0.8 m²
6.	Fix ply formwork to beam and deck soffits	0.6 m²
7.	Fix and strike starter forms to walls	6.0 m
8.	Fix and strike starter forms to columns	2.5 m
9.	Fix and strike stop ends for expansion joints	0.7 m²
10.	Strike, clean, repair, and stack formwork for reuse	1.7 m²
11.	Make, fix, and strike sundry formwork in boxes, pits, plinths, bearing pads, and so on	0.4 m²

4.11   PRODUCT MANAGEMENT

The typical life cycle of a product used in civil engineering generally corresponds to the cycles shown in Figure 4.8. Initially, during the development, introduction, and growth stages, the amount of work expended on a product is proportionately high. The turnover is likely to reach a maximum somewhere in the early maturity stage, whereas the

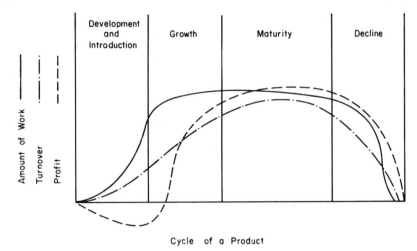

Cycle of a Product

Figure 4.8   Life cycle of a product.

profit is likely to be negative initially, that is, a loss, and to be at a maximum toward the end of the maturity stage.

The cost of making a product or even building a house depends to a certain extent on the number of repetitions. The unit costs are reduced in a manner similar to that shown in Figure 4.9, but the actual percentage reduction depends on the type of product.

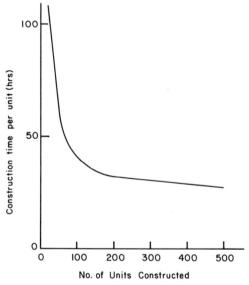

Figure 4.9   Construction time per unit versus number of units.

4.12 WORK STUDY

Work study is the systematic examination of any job so that the most effective and economical use of labor, material, and equipment resources can be obtained. Work study is frequently split into two divisions, called method study and work measurement. These divisions are defined in Figure 4.10.

Work measurement is a relatively simple exercise of measurement and timing, although the definition of a fair work load is difficult. For instance, as time progresses an identical activity on site should take less time for completion.

Method study is usually divided into several chronological stages. The first stage is where the method that has been assigned or selected is examined for improvement. Items for detailed study should include the possibility of mechanization, unnecessary repetition, bottlenecks, danger, time consuming activities, physically tiring manual jobs, and so on.

A job can be improved more easily if each activity is isolated and studied separately. The job is therefore recorded step by step on a flow process chart similar to the chart shown in Figure 4.11. A symbol is frequently placed adjacent to each activity as shown on the chart. The symbols represent the following:

$\bigcirc$ Operation
$\square$ Inspection
$\Rightarrow$ Transportation
D Delay
∇ Storage

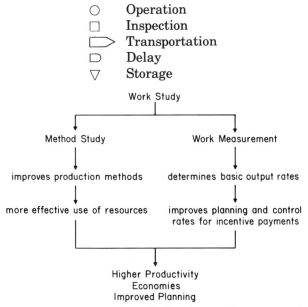

Figure 4.10 Work study/method study—work measurement.

Step Description	Quantity	Distance	Time	Operation	Inspection	Transport	Storage	Delay	Remarks
Steel reinforcement delivered						▷			By rail/road
Unload			75	O					Mobile crane
Sorted			20	O					By hand
To storage racks		70				▷			Mobile crane
Storage							▽		
Await delivery to cutting bench								D	
To cutting bench		60				▷			Mobile crane
Cut to length			110	O					3 steelworkers
Await delivery to bending bench								D	
To bending bench		10				▷			
Steel bent			360	O					3 steelworkers
To site		250				▷			
Await fixing								D	Blinding not ready
Fix steel			700	O					4 steelworkers
Contractor's Engineer inspection					▢				Engineer/ Foreman
Cleaning out			30	O					
Await R.E. inspection								D	
R.E. inspection					▢				
Concrete poured			180	O					8 Concrete gang
Simon O'Neil 1981-6-3									

Figure 4.11 A flow process chart.

A flow diagram, as shown in Figure 4.12, is frequently used to show the movement of a consignment of material.

Each step in the method is then examined very carefully for possible alteration. This leads to the development of the most practically efficient and economical method. Throughout the study everyone connected to the job should be asked for ideas and suggestions.

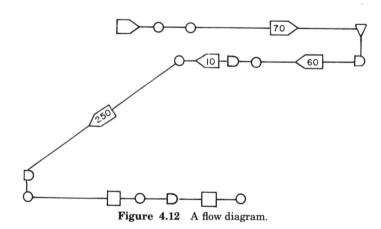

Figure 4.12 A flow diagram.

If after the study a new method is applied, it should be subjected to the same rigorous study as the former method. Everyone connected to the new method should be clearly informed of the alterations before it is applied. Further training may be necessary in some cases.

Finally the work study engineer should return to check that the new method has achieved the desired objectives after it has been operating for some time.

4.13 WORKED EXAMPLES

Note that Examples 4.1 to 4.6 have been inserted at specific points within the text and not at the end of the chapter to facilitate the learning process and maintain continuity.

Example 4.7

A machine required in a construction project requires an initial outlay of $1,000,000. Forecasts indicate that the machine is capable of recouping $300,000 p.a. for the first 3 years and $200,000 p.a. for the following 2 years. Show that the machine is economically viable if it is assumed that the machine has a life of 5 years and that a minimum dollar return of 10% p.a. is required.

Answer 4.7

At a discounting of 10% the present value is shown in the following table:

(1) At year end	(2) Equivalent cash receipts	(3) $(1 + r)^n$	(4) PV (2) $\div$ (3)
0	−1000	1	−1000
1	300	1.10	272.7
2	300	1.21	247.9
3	300	1.331	225.4
4	200	1.4641	136.6
5	200	1.6105	124.2
NPV			6.8

Values in $1000.

As the total NPV is positive, the machine may be considered to be economically viable. However, the required return on investment of 10% is a fictitious percentage if inflation is not negligible.

Example 4.8

If the rate of inflation is 5%, express the cash flow in Example 4.7, before discounting, in real terms.

Answer 4.8

(a) Year end	0	1	2	3	4	5
(b) Dollar cash flow	−1000	300	300	300	200	200
(c) Compounded inflation rate	1.00	1.05	1.1025	1.1576	1.2155	1.2763
(d) Real cash flow (b)/(c)	−1000	285.7	272.1	259.2	164.5	156.7

Values in $1000.

Example 4.9

If the rate of inflation is 5% and a dollar return on investment of 10% (i.e., fictitious) is required, what is the required return in real terms?

Answer 4.9

The real rate of return can be deduced from the following equation:

$$(1 + r_a) = (1 + r)(1 + r_i) \text{ (equation 4.6, Section 4.5)}$$

$$r_a = \text{apparent or fictitious interest rate}$$
$$r_i = \text{inflation rate}$$
$$r = \text{real interest rate}$$

$$1 + r = \frac{1.10}{1.05} = 1.0476$$

real rate of return = 4.76%

Example 4.10

By comparing the results with the data obtained in Example 4.7, use the following table to show that the net DCF is the same for a real rate of return of 4.76% compared with a 10% apparent return on investment and a 5% inflation rate.

Real cash flow	−1000	285.7	272.1	259.2	164.5	156.7
Year	0	1	2	3	4	5

Values in $1000.

Answer 4.10

Year	0	1	2	3	4	5
Real cash flow	−1000	285.7	272.1	259.2	164.5	156.7
4.76% real rate	1.0	1.0476	1.0975	1.1498	1.2045	1.2618
Real DCF +	−1000	272.7	247.9	225.4	136.6	124.2
Real net DCF	−1000	−727.3	−479.4	−254.0	−117.4	6.8

In $1000 units.

It can be seen that the data in this answer directly compares with the data in Answer 4.7.

Example 4.11

If a loan of $1 million is required over a period of 5 years, deduce the annual required repayments using a method where annual equal repayments are made on the capital as well as interest repayments on the outstanding capital (i.e., the straight line redemption method). Assume an interest rate of 6%.

Answer 4.11

Year end	(a) Opening balance	(b) Equal repayments	(c) Remaining balance	(d) Interest on opening balance	(e) = (b) + (d) Annual repayment
1	1000	200	800	60	260
2	800	200	600	48	248
3	600	200	400	36	236
4	400	200	200	24	224
5	200	200	0	12	212
				Total repayment	1180

In $1000 units.

Example 4.12

Using the same data as in Example 4.11, use the method of equal annual payments to determine the total annual payment and the closing balance at the end of each year. Refer to equation 4.11, Section 4.7.

Answer 4.12

$$X = \frac{Ar}{1 - 1/(1 + r)^n} \text{ (from equation 4.11)}$$

$$= \frac{1,000,000 \times 0.06}{1 - 1/(1 + 0.06)^5}$$

$$= \frac{1,000,000 \times 0.06}{1 - 0.74726}$$

$$= \$237,396.40$$

Year	(a) Opening balance	(b) Interest on balance	(c) Total annual payment	(d) = (c) − (b) Repayment on balance at year end	(e) = (a) − (d) Closing balance
1	1000	60	237.4	177.4	822.6
2	822.6	49.4	237.4	188.0	634.6
3	634.6	38.1	237.4	199.3	435.3
4	435.3	26.1	237.4	211.3	224
5	224	13.4	237.4	224	0
		Total repayment	1187		

In $1000 units.

Example 4.13

The data shown below were recorded on a civil engineering project. In this particular case the term "cost" means the actual payments of costs by the contractor, instead of the more usual "cost incurred" meaning, where a time lag on payment of materials may be assumed.

Month end	Value of work completed	Monthly payment: owner to contractor	Costs paid by contractor
1	60	20	30
2	90	30	50
3	50	50	70
4	170	60	120
5	130	100	80
6	200	220	150
7	90	220	100
8	110	150	140
9	120	140	70
10	40	30	110
11	60	30	20
12	30	20	30
13		30	30
Retention		50	

Values in nearest $1000.

At the beginning of the contract it was anticipated that $220,000 would need to be borrowed to assist the cash flow at the most critical time.

(a) Determine if a loan of $220,000 was sufficient
(b) How much money did the owner owe the contractor at the critical cash flow time?
(c) Show the amount of profit made by the contractor
(d) What amount of loan did the contractor require to finance the project up to the end of the third month?
(e) When could the contractor pay back all the loan?

Answer 4.13

Month end	Value (monthly cumulative)		Payments (monthly cumulative)		Costs (monthly cumulative)	
1	60	60	20	20	30	30
2	90	150	30	50	50	80
3	50	200	50	100	70	150
4	170	370	60	160	120	270
5	130	500	100	260	80	350
6	200	700	220	480	150	500
7	90	790	220	700	100	600
8	110	900	150	850	140	740
9	120	1020	140	990	70	810
10	40	1060	30	1020	110	920
11	60	1120	30	1050	20	940
12	30	1150	20	1070	30	970
13			30	1000	30	1000
Retention			50	1150		

Values in nearest $1000.

The cumulative values in the above table are plotted in Figure 4.13 to simplify the solutions.

(a) A loan of $220,000 was not sufficient; $240,000 was required
(b) The owner owed the contractor $440,000 at the critical cash flow time
(c) The contractor made $150,000 profit
(d) The contractor required a $100,000 loan over the first 3 months of the contract
(e) The contractor did not need the loan after the end of the eighth month, as the cumulative costs did not subsequently exceed the cumulative payments

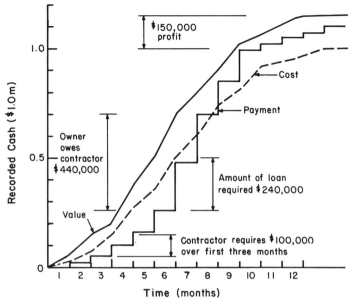

Figure 4.13 Recorded cash versus time.

Example 4.14

One chargehand and three steel fixers cut and bent the following amount of steel reinforcement in 1 week:

cut: 25 mm diam.,   850 kg   bent: 25 mm diam.,   850 kg
     35 mm diam., 10,600 kg   32 mm diam., 10,600 kg
                                      40 mm diam., 20,875 kg

Deduce

(a) The number of hours saved (i.e., target hours less actual hours)
(b) The bonus for each operative if

CH worked 55 hr (basic rate $11/hr)
SF1 worked 55 hr (basic rate $10/hr)
SF2 worked 50 hr (basic rate $10/hr)
SF3 worked 45 hr (basic rate $10/hr)

Answer 4.14

(a)

	Diameter (mm)	Weight (kg)	Target (kg/hr)	Target hours
Cut	25	850	300	2.8
	32	10,600	400	26.5
Bent	25	850	100	8.5
	32	10,600	125	84.8
	40	20,875	125	167.0

Total number of target hours = 289.6

$$\text{Actual hours worked by CH} + 3\text{SF} = 205$$
$$\text{Total number of target hours} = 290$$
$$\text{Hours saved} = 85$$

(b) Average basic rate = 10.25

Total bonus $= 10.25 \times 85 \times \dfrac{50}{100}$

$= \$435$

Individual bonus (CH, for example) $= \dfrac{605}{2105} \times 435$

$= \$125$ (refer to share values in table)

Operative	Hours worked	Basic rate ($)	Share value	Bonus ($)
CH	55	11	605	125
SF1	55	10	550	114
SF2	50	10	500	103
SF3	45	10	450	93
			2105	$435

CHAPTER 5

FACTORS INFLUENCING
EQUIPMENT PERFORMANCE

5.1 TYPES OF EQUIPMENT

The types of equipment used on a civil engineering site may be broadly categorized under the job descriptions shown in Table 5.1. The selection, use, and purpose of equipment for different job types are discussed in Chapters 6 to 12. More detailed information can be obtained from manufacturers of the equipment.

Before the selection and performance of equipment on a construction site are considered, the soil properties should be examined. For instance, the change of volume of a soil or rock in its in situ, loose, or compacted state is a very important factor in earth moving operations.

5.2 UNIT WEIGHT, BULKING, COMPACTION, AND VOLUME CAPACITY

The in situ unit weight of a certain type of soil is likely to vary appreciably depending on whether it is dry or saturated. Approximate values for the unit weight of certain soil-rock types are given in Table 5.2. The in situ unit weight is often referred to as the undisturbed or bank density.

When a soil is excavated, the number and volume of the voids increase, resulting in an increase in volume with a corresponding decrease in unit weight. This change in volume is known as bulking or swelling.

If the coefficient of bulking C_b is expressed as an increase in volume, then

$$V_l = (1 + C_b)V_i \qquad (5.1)$$

TABLE 5.1 Job Description—Types of Equipment

Lifting	Excavating
Crawler mounted crane	Face shovel
Truck mounted crane	Dragline
Self-propelled mobile crane	Grab
Tower crane	Hydraulic excavator
Forklift	Excavator-loader
Derrick	Front end loader
	Trenching machine

Excavating/Transporting	Transporting
Tractor drawn scraper	Dump truck
Rubber tired scraper	Dumper
Bulldozer and angledozer	Belt conveyor
Grader	Monorail, cable way, and so on

Compaction	Tunneling
Smooth wheel roller	Shield
Pneumatic tired roller	Tunneling machine in soft ground
Grid roller	Tunneling machine in rock
Sheepsfoot and tamping rollers	Bentonite machine
Vibrating roller	
Small compaction machines	

or

$$C_b = \frac{V_l}{V_i} - 1 \tag{5.2}$$

where V_l = loose volume
          V_i = in situ volume

For a given weight of soil the unit weight γ is inversely proportional to the volume. Therefore

$$C_b = \frac{\gamma_i}{\gamma_l} - 1 \tag{5.3}$$

where γ_i = in situ unit weight (bulk density)
          γ_l = loose unit weight

The coefficient of bulking is likely to vary for a certain type of soil because of the physical variation in the soil or the manner in which it is excavated. For example, the coefficient of bulking varies depending on whether the soil has been dug or scraped.

TABLE 5.2 Soil Types—Unit Weight and Coefficient of Bulking

Soil-rock type	In situ unit weight (kN/m³)		Coefficient of bulking
	Dry-moist	Saturated	
Rock	20 to 27	20 to 27	0.30 to 0.70
Gravelly soils	15 to 22	17 to 24	0.10 to 0.20
Sandy soils	14 to 21	15 to 23	0.05 to 0.15
Sands-silts (LC)	14 to 17	15 to 19	0.20 to 0.40
Sands-silts-clays	14 to 18	15 to 19	0.30 to 0.60
Organic soils	< 17	< 17	0.10 to 0.25

LC = low compressibility.

Table 5.2 shows the ranges of values for the coefficient of bulking for various soil-rock types. Because of the variable nature of soil, calculations involving the bulking coefficient cannot be too precise until the soil has been exposed and excavated. It would therefore be fairer and more precise if the price quoted at the tender stage by the contractor, in large earth moving operations, contained price adjustment formulae to allow for a variation in the coefficient of bulking.

A given weight of soil, when properly compacted, can occupy a volume that is less than the original in situ volume because of the expulsion of air from the soil.

If the coefficient of compaction C_c is expressed as a decrease in volume, then

$$V_c = (1 - C_c)V_i \tag{5.4}$$

where V_c = compacted volume

or

$$C_c = \left(1 - \frac{V_c}{V_i}\right) \tag{5.5}$$

and

$$C_c = \left(1 - \frac{\gamma_i}{\gamma_c}\right) \tag{5.6}$$

where γ_c = compacted unit weight.

Table 5.3 shows the range of values for the coefficient of compaction. The value of the soil type as a road foundation progressively decreases down the list in the table.

TABLE 5.3   Soil Types—Coefficient of Compaction

Soil-rock type	In situ dry unit weight (kN/m^3)	Coefficient of compaction: no compaction/proper compaction
Rock	20 to 27	$- 0.70$ to 0.00
Gravelly soils	15 to 22	$- 0.20$ to 0.20
Sandy soils	14 to 21	$- 0.15$ to 0.22
Sands-silts (LC)	14 to 17	$- 0.40$ to 0.05
Sands-silts-clays	14 to 18	$- 0.60$ to HC
Organic soils	< 17	$- 0.25$ to HC

HC = soils possessing high compressibility; LC = soils possessing low compressibility.

The unit weight of a soil or rock can vary with the degree of compaction, the moisture content, and the texture and size of the rock pieces. If loose soil is deposited in an area as fill material without compaction, it possesses many unsuitable voids, because compaction equipment reduces the number and size of the voids by expelling air and moisture by vibration or pressure. It is worth noting that both the coefficient of bulking and the coefficient of compaction refer to the in situ volume of soil. The same comment for the inclusion of price adjustment formulas may be made concerning the coefficient of compaction as was previously made for variations in the coefficient of bulking.

The optimum volume carried by a scraper may be only about 90% of the volume corresponding to a full load because the extra time and tractive effort to fully load a scraper is generally counterproductive. However, if the scraper is pushed, it is usual to assume that the scraper carries a full load.

The volume carried by earth moving equipment depends on whether the load is heaped or level, the type of soil controls at which angle the soil may be heaped without spillage. Additional end and side boards can increase the volume capacity of a truck, but the maximum volume carried should not exceed the equivalent of the rated load.

5.3   POWER CONSIDERATIONS: ROLLING RESISTANCE, GRADIENT, TRACTION, ALTITUDE, AND MINIMIZING POWER REQUIREMENTS

Power is provided by the engine or prime mover in construction equipment. The available power is the force from the power unit applied at the rim of the wheels or base of the tracks, in contact with the soil. If the power available is not sufficient, the load carried can be reduced or a machine with more horsepower can be used. Available power depends on the horsepower divided by the speed and therefore increases as the speed of the machine decreases. The power required equals the rolling resistance plus the grade resistance. Both resistances can be measured in equivalent kilonewtons. It is convenient for civil engineers to consider only external factors that resist the movement of earth moving equipment on site. Internal factors, such as wheel bearing friction, may be subsumed in a factor allowing for the rolling resistance of the vehicle. Refer to Figure 5.1*a, b*.

In plant manufacturers' literature the power available is known as rimpull when it refers to the force exerted by the rubber tires at the rim of the wheels at the ground contact point, or drawbar pull when it refers to the force in the drawbar between the tracked prime mover and the trailing vehicle.

Equipment requires less power to travel over a hard surface than over soft ground, and requires more power the heavier the load.

Rolling resistance is the resistance to movement of plant over level ground and may be expressed as a percentage of the total weight of the machine and load. Refer to Table 5.4. Part of the rolling resistance increases in direct proportion to the depth of the tire penetration in the soil and the other part depends on the tire flexure.

For tracked vehicles only the values quoted for the drawbar pull include an allowance for the very small amount of rolling resistance, which means that in earth moving calculations the rolling resistance need not be considered. For combined tracked and wheeled vehicles,

TABLE 5.4   Rolling Resistances for Various Surfaces

Surface	Rolling resistance (RR) %	
	Tracked and wheeled combination	Rubber tired
Hard road	2.75 ± 0.25	2.00 ± 0.25
Flexible road	3.25 ± 0.50	2.50 ± 0.75
Well maintained, compacted haul road	3.50 ± 1.00	2.75 ± 0.75
Badly maintained haul road	7.00 ± 3.00	7.50 ± 3.50
Loose sand and/or gravel	9.00 ± 1.00	13.00 ± 2.00
Soft, muddy haul road	11.00 ± 1.00	17.00 ± 3.00

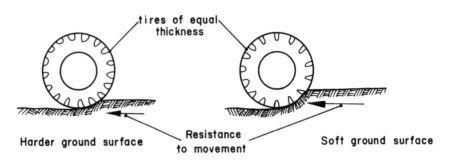

Less rolling resistance

More rolling resistance

tires of equal thickness

Harder ground surface

Resistance to movement

Soft ground surface

(a)

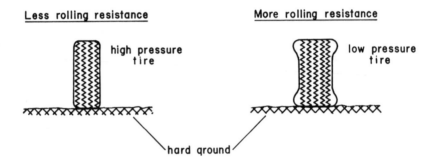

Less rolling resistance

More rolling resistance

high pressure tire

low pressure tire

hard ground

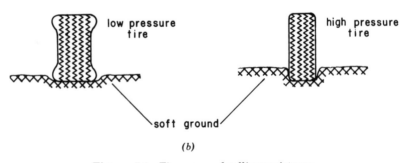

low pressure tire

high pressure tire

soft ground

(b)

Figure 5.1 Tire—ground rolling resistance.

such as a tractor drawn scraper, the power required to overcome the rolling resistance should be calculated with reference to the total weight on the wheeled unit only.

For a given hard surface the higher range of values in Table 5.4 apply to tires with lower air pressure, which occupy a greater contact area and thus produce a greater rolling resistance. However, the opposite is applicable for the loose and soft ground surfaces, where the higher value ranges of the rolling resistance are appropriate to tires with higher pressures because of increased penetration into the ground.

The power required to move machines uphill is greater than that required over level ground. Conversely, use should be made of the power advantage for machines moving downhill whenever possible.

The increase in power required to keep the machine moving is the force component of the weight of the machine M_G parallel to the sloping ground surface, see Figure 5.2.

$$\frac{M_G}{M} = \frac{1}{\sqrt{1 + n^2}}$$

where M is the weight of the machine and M_p is the force component acting perpendicularly to the ground surface.

Table 5.5 shows that it is sufficiently accurate, for the average slopes encountered on site, to assume that

$$\frac{M_G}{M} = \frac{1}{n} = \frac{G}{100}$$

there being only a 3% error in a slope of 1 in 4, or a 25% gradient.

The resistance to skidding or slipping between the wheels or tracks of a vehicle and the contact area of the ground is called traction. The maximum force that a machine can deliver may be limited by traction. Traction depends on the weight on the drive tires or tracks, the type of

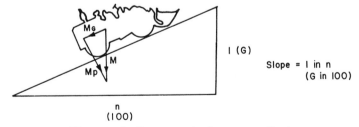

Figure 5.2 Power required versus gradient.

TABLE 5.5 Effect of
Power on Gradient

n	G	M_G/M
100	1	0.0100
50	2	0.0200
25	4	0.0400
20	5	0.0499
10	10	0.0995
6.67	15	0.1483
5	20	0.1961
4	25	0.2425

haul road surface, and the pressure in the tires or the width of the tires or tracks. The maximum power delivered may therefore be limited either by traction or the maximum power provided by the prime mover in first gear.

The coefficient of traction (Figure 5.3), when multiplied by the total load on the tracks or drive tires, gives the maximum tractive force that can be exerted between the tire or track and the ground before slipping occurs. To determine this tractive force the following equation can be used:

$$P = W \tan \phi$$

where $\tan \phi$ = coefficient of traction
W = weight on drive tires
ϕ = limiting angle of resistance
P = forces available to move equipment, before slipping occurs

It can be seen in Table 5.6 that traction for a wheeled machine is

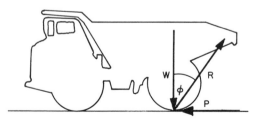

Figure 5.3 Coefficient of traction.

TABLE 5.6   Coefficient of Traction, tan ϕ

Type of surface	Tracked plant	Wheeled plant
Concrete	0.45	0.80 to 0.95
Asphalt	n.a.	0.65 to 0.70
Dry clay	0.85 to 0.90	0.50 to 0.70
Wet clay	0.70 to 0.85	0.40 to 0.50
Dry sandy ground	0.30	0.20 to 0.35
Wet sandy ground	0.35 to 0.50	0.35 to 0.40

better than a tracked machine over a hard smooth surface but generally poorer over ground such as clay.

On most surfaces except sandy soils traction deteriorates when rain occurs and the coefficient of traction becomes lower as the ground surface becomes wetter.

The decrease in oxygen in the air at high altitudes affects the performance of earth moving machines. Specification sheets show how much available power a machine can produce in a given gear over a certain range of speed when the engine is operating at its rated horsepower. At very high altitudes the decreased density of the air produces a corresponding decrease in rimpull or drawbar pull. Most machines operate up to an elevation of about 1500 m before they begin to lose any appreciable power due to the rarity of the air, and some turbocharged engines can operate up to an elevation of 3000 m before they require derating. Usually the effect of altitude can be considered by increasing the appropriate time components of the cycle time.

Calculations involving the production of earth moving equipment are by their very nature inexact. For example, a shower of rain can alter the performance of a machine. However, although the production estimates may not be precise, it is possible to greatly reduce the power requirements of equipment on site.

One of the most important factors is the haul road surface. Motor graders and sprinkler facilities can greatly improve conditions for both production and safety reasons. A good layer of hardcore over the haul road surface may often be very economical on overall costs.

When possible, where fill is to be both imported and transported on site, use should be made of the help of the average gradient by hauling the load downhill and returning empty uphill. The areas of cut and fill are likely to alter as work progresses on a large earth moving operation, and therefore every effort should be made to ensure that the fleet of scrapers travels on the straightest and shortest routes. When the soil

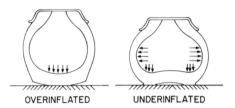

OVERINFLATED   UNDERINFLATED

Figure 5.4 Effect of incorrect tire pressures.

is actually being loaded, a machine that pushes the scrapers reduces the time of loading.

The correct tire pressure and correct type and width of tracks also help decrease the power required to haul earth. Incorrect tire pressures (Figure 5.4) not only reduce the power available, but are dangerous and can very significantly reduce the life of the tires.

5.4   CYCLE TIMES, FIXED TIME, VARIABLE TIME, AND OPERATING EFFICIENCY

The time taken for earth moving machines to load, haul, and dump material and then return empty to the starting position ready for the second load is known as the cycle time. This cycle is diagrammatically shown in Figure 5.5. The cycle can be divided into sections. Once the machine is operating in the appropriate gear, the velocity is considered to be constant. The times for the haul and return journeys depend on the distance traveled and therefore vary considerably. Specification sheets are available which give the fixed times on the variable velocity loading and dumping sections.

The fixed time constants depend on the type of machine and on whether the scraper is single engined, in tandem, push loaded or self-loaded. Differences in machines, such as all-wheel drive machines and elevating scrapers, affect the fixed time values, so it is essential to consult the machine manufacturers' specification sheets and handbooks.

Figure 5.5 Diagrammatic sketch of earth moving equipment cycle.

The large multinational companies produce very informative manuals and handbooks in which all the variables on a particular site can be considered so that the cycle time can be assessed fairly accurately. Table 5.7 shows fixed times for various operations and types of scraper. It should again be emphasized that these values may not be precise because of the very nature of civil engineering earth moving operations.

A graph of available rimpull in relation to the gears of a wheeled scraper is shown in Figure 5.6. Charts supplied by the plant manufacturer, similar to Figure 5.6, enable the planning engineer to deduce the value of the maximum rimpull available for a machine in a given gear.

The power required is estimated for the haul journey when the scraper is loaded, and this value can be used to deduce in which gear, and consequently at what speed, the scraper should travel. By knowing the length of the haul route, the time taken for hauling can be determined. A similar calculation can be made for the return journey when the scraper is unloaded. It should be noted that, in practice, the haul distance can vary over the weeks of operation for the same cut and fill areas, but an average distance obtained from the mass haul diagram is

TABLE 5.7 Fixed Times for Scrapers

	Fixed time (sec)			
	Tracked-wheeled scraper		Wheeled scraper	
	Self-loading	Pushed	Self-loading	Pushed
Loading	90	60	55–60	35[a]
Dumping and maneuvering	60	60	40	40
Acceleration and de-celeration[c]	0	0	60[b]	60[b]

[a]It is assumed for this time that one pusher is used for machines up to about 20 m³ heaped capacity and that two pushers are used for machines over 20 m³ capacity.

[b]This time may be taken for plant traveling at a haul speed of 8 m/sec (30 km/hr). Approximately 10 sec per 1 m/sec (3 sec per 1 km/hr) should be added or subtracted for different haul speeds.

[c]It should be noted that some machine manufacturers who produce charts to determine the haul and return travel times include the machine acceleration and deceleration times within the travel times.

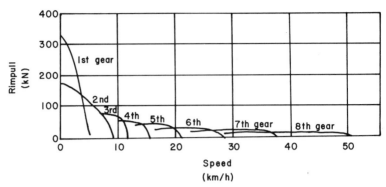

Figure 5.6   Rimpull versus gears—wheeled scraper.

frequently assumed. Over prolonged periods of work in the same area, however, different distances should be calculated and used.

Some engines contain a torque converter drive that automates the manual gear changes in the normal direct drive transmission.

The number of minutes actually worked per hour may be determined by a work study engineer. This value is likely to vary from one site to another for a variety of reasons. An average value usually assumed ranges between 45 to 50 minutes per hour operating efficiency. The predicted cycle time should be increased to allow for this operating efficiency.

Many parameters govern the choice of equipment for bulk earth moving operations. All sites vary due to the differences in soil, haul distance, and gradient. A very general indication of optimum economic haul distances for certain machines is as follows:

0 to 100 m	Dozer
100 to 500 m	Tractor drawn scraper
300 m to 1.5 km to 2.0 km	Rubber tired scrapper
Over 1.5 km	Trucks and front-end loaders or shovels
	Conveyors

An essential task for the estimator is to determine the unit hourly cost of equipment. The hourly cost, apart from the operator's cost, can be computed from the following items:

1. Depreciation
2. Mechanical repairs and spares
3. Tires, if applicable
4. Fuel
5. Interest, insurance, and taxes
6. Oils and grease

These items are discussed in more detail in Chapter 10.

The weather can greatly affect the performance of equipment on site. On wet soil surfaces traction may be greatly reduced and rolling resistance may be increased. A variation in temperature can also affect output. Earthworks should be planned, if possible, so that they can be carried out in the most favorable seasons.

5.5 WORKED EXAMPLES

Example 5.1

Determine the coefficient of bulking and the coefficient of compaction for the following soil. The unit weights are as follows:

In situ	15 kN/m³
Loose	12 kN/m³
Compacted	17 kN/m³

Which unit weight should the estimating department in a civil engineering company use to determine costs for the excavation, transportation, and fill items?

Answer 5.1

Using equation 5.3, we find the coefficient of bulking:

$$C_b = \left(\frac{\gamma_i}{\gamma_l} - 1 \right)$$

$$= \left(\frac{15}{12} - 1 \right)$$

$$= 0.25$$

By equation 5.6 the coefficient of compaction is

$$C_c = \left(1 - \frac{\gamma_i}{\gamma_c}\right)$$

$$= \left(1 - \frac{15}{17}\right)$$

$$= 0.118$$

The calculations involved in:
(a) The excavation item should use 15 kN/m³
(b) Transportation should use 12 kN/m³
(c) The fill item should use 17 kN/m³

Example 5.2

(b) A wheel tractor scraper has a capacity of 40 m³ when the soil is heaped. If the coefficient of bulking is 0.20 and the in situ unit weight of the soil to be excavated is 20 kN/m³, determine the rimpull requirements of the scraper for the haul and return journey if the weight of the tractor scraper, when empty, is 333 kN. Assume the following:

Haul journey	For 1.5 km, level, and 0.5 km, 1 in 10 downhill
Return journey	For 2 km, 1 in 30 uphill
Rolling resistance	3.50%

Answer 5.2

By equation 5.1 the loose volume is

$$V_l = (1 + C_b)V_i$$

$$V_i = \frac{40}{1 + 0.2} = 33.3 \text{ m}^3$$

$$\text{load carried} = 33.3 \times 20$$

$$= 667 \text{ kN}$$

Rimpull requirements:
(a) For haul journey, 1.5 km level,

$$\text{Rolling resistance} = (333 + 667) \times 3.5/100 = 35 \text{ kN}$$

(b) For haul journey, 0.5 km, 1 in 10 downhill,

Rolling resistance = 35 kN

$$\text{Grade assistance} = (333 + 667) \times \frac{10}{100} = \underline{100 \text{ kN}}$$
$$\underline{- 65 \text{ kN}}$$

Therefore no rimpull is required.

(c) For return journey, 2.0 km, 1 in 30 uphill,

$$\text{Rolling resistance} = 333 \times \frac{3.5}{100} = 11.6 \text{ kN}$$

$$\text{Grade resistance} = 333 \times \frac{1}{30} = 11.1 \text{ kN}$$

Example 5.3

If a truck has a 200 kN load on its drive wheels and the coefficient of traction for the surface material is 0.5, at what rimpull do the wheels begin to slip? Comment if the maximum amount of power required is 120 kN.

Answer 5.3

The wheels begin to slip at $200 \times 0.5 = 100$ kN rimpull. If the rimpull requirements are 120 kN, the truck is not able to move because the drive wheels slip.

Example 5.4

Iron ore is to be excavated with a face shovel and hauled away by trucks. Four different sizes of trucks are available for the operation. The payloads for the truck are as follows:

Truck type	Payload (kN)
Unit 1	500
Unit 2	635
Unit 3	770
Unit 4	1360

The bucket capacity for the shovel is 5.36 m³ and the unit weight of the material is 48 kN/m³. If the unit is to be loaded in three or four passes of the shovel, which unit should be recommended?

Answer 5.4

The payload for three passes of the shovel is

$$3 \times 5.36 \times 48 = 772 \text{ kN}$$

The payload for four passes of the shovel is

$$4 \times 5.36 \times 48 = 1029 \text{ kN}$$

Unit 3 is therefore the most suitable truck to use with three passes of the face shovel.

Example 5.5

A contract requires 700,000 m³ of compacted fill, and 450,000 m³ of suitable undistorbed soil is available on site.
   The imported fill has a 20% bulking factor and a 17% compaction factor, and the suitable in situ soil, within the contract limits, has a 20% bulking factor and a 15% compaction factor.

Determine
(a)   the total number of cycles required by scrapers to haul the in situ soil
(b)   the total number of trips for trucks hauling the imported fill

The capacity of the scrapers may be assumed to be 30 m³ and that of the trucks 25 m³.

Answer 5.5

700,000 m³ is required.
The scrapers are to haul 450,000 m³ of the in situ soil:

$$C_b = (V_l/V_i - 1) \text{ (see equation 5.2)}$$

$$\frac{20}{100} = \frac{V_l}{V_i} - 1$$

$$V_l = 1.2 \times V_i$$
$$= 1.2 \times 450,000$$
$$= 540,000 \text{ m}^3 \text{ loose volume}$$

$$\text{No. of scraper cycles} = \frac{540,000}{30}$$

$$= 18,000$$

The scrapers therefore haul 450,000 m³ of the in situ soil. The equivalent compacted volume should now be determined:

$$C_c = 1 - \frac{V_c}{V_i} \text{ (see equation 5.5)}$$

$$\frac{15}{100} = 1 - \frac{V_c}{450,000}$$

$$V_c = 0.85 \times 450,000$$
$$= 382,500 \text{ m}^3 \text{ compacted}$$

The trucks are therefore required to haul 700,000 − 382,500 = 317,500 m³ compacted. It is necessary to convert this compacted volume to a loose volume to determine the number of trips required by the trucks.

By combining equation 5.1,

$$V_l = (1 + C_b)V_i$$

and equation 5.4,

$$V_c = (1 - C_c)V_i$$

$$V_l = \frac{V_c}{(1 - C_c)}(1 + C_b)$$

$$= \frac{317,500}{(1 - 17/100)}\left(1 + \frac{20}{100}\right)$$

$$= \frac{317,500 \times 1.2}{0.83}$$

$$= 459,036 \text{ m}^3$$

$$\text{No. of truck trips} = \frac{459,036}{25}$$

$$= 18,361$$

Example 5.6

Determine the cycle time for a rubber tired tractor scraper with a maximum capacity of 20 m³ for a project having the following data:

(a) The average haul distance is 3.0 km down an average gradient of 2 in 100. The return journey should be assumed to be the reverse.
(b) The material to be hauled weighs 18 kN/m³ in the natural undisturbed state and has a bulking coefficient of 0.15.

(c)　The estimated rolling resistance is 5%.

(d)　The efficiency of the scraper is 75%.

Scraper specification

(a)　Because of the doubtful nature of the soil it is intended to push the tractor scraper with another tractor during the loading operation. The loading time for the scraper should be taken as 1.00 min and the dumping and turning time as 1.00 min.

(b)　The weight of the tractor and scraper is 300 kN, and the weight on the drive wheels when loaded is 180 kN.

(c)

Gear	Speed (m/sec)	Available rimpull (kN)
1st	2	100
2nd	4	65
3rd	8	40
4th	12	20
5th	20	12

Answer 5.6

Scraper capacity $= 20$ m³ loose

$$V_l = (1 + C_b)V_i \text{ (equation 5.1)}$$

$$V_i = \frac{V_l}{(1 + C_b)}$$

$$= \frac{20}{1.15}$$

$$= 17.39 \text{ m}^3 \text{ in situ}$$

Total load/scraper/cycle $= 17.39 \times 18 = 313$ kN

Power requirements:

(a)　Haul

$$\text{Rolling resistance} = (313 + 300) \times \frac{5}{100} = 30.65 \text{ kN}$$

$$\text{Grade assistance} = 613 \times \frac{2}{100} \qquad = \underline{12.26 \text{ kN}}$$
$$\text{Net power required} \qquad\qquad\qquad 18.39 \text{ kN}$$

Haul, 3.0 km, 18.39 kN rimpull required, hence 4th gear.

(b) Return

$$\text{Rolling resistance} = 300 \times \frac{5}{100} = 15 \text{ kN}$$

$$\text{Grade resistance} = 300 \times \frac{2}{100} = \underline{\ 6 \text{ kN}}$$

$$\qquad\qquad \text{Total power required} \qquad 21 \text{ kN}$$

Return, 3.0 km, 21 kN rimpull required, hence 3rd gear.

Cycle time, given D9 dozer pushing:

$$\text{Fixed time, given} \qquad = \ 2.00 \text{ min}$$

$$\text{Haul time } \frac{3000}{12} \times 60 \quad = \ 4.17 \text{ min}$$

$$\text{Return time } \frac{3000}{8} \times 60 = \ 6.25 \text{ min}$$

$$\qquad \text{Cycle time} \qquad\qquad \overline{12.42 \text{ min}}$$

75% efficient

$$\qquad \text{Actual cycle time} \quad = \frac{12.42}{0.75} \text{ min}$$

$$\qquad\qquad\qquad\qquad\qquad = 16.56 \text{ min}$$

CHAPTER 6

LIFTING EQUIPMENT

6.1 DEFINITIONS

A lifting crane is a machine of the type shown in Figure 6.1, operated by a single or multipart rope. The horizontal or inclined member from which the load is suspended is called the jib or the boom. Other terms used when describing lifting or associated equipment are as follows:

Ballast (kentledge)	Dead weight added to a crane to ensure stability
Counterweight	Weights added to a crane, often at the back of the cab to provide a counterbalancing effect
Fly jib	A jib attached to the top end of the main jib to increase its outreach or height
Luffing	The action of moving the jib in the vertical plane about the jib pivot
Outreach	The horizontal distance from the crane hook to the nearest point of the cab
Outriggers	Steadying retractable arms extending from the crane chassis onto the ground to provide additional stability
Safe working load (rated load, rating)	The maximum safe load that, under specified conditions, can be safely handled
Slewing	The action of moving the jib, and usually the cab, in the horizontal plane

The stability of a crane depends on the weight of the crane, which acts as a counterweight to the load lifted, hence preventing the machine from tipping over. Refer to Figure 6.1.

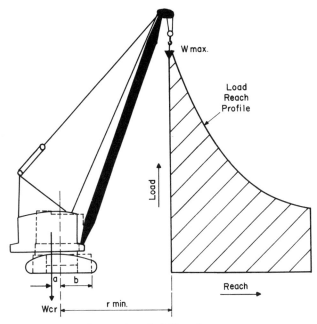

Figure 6.1 Stability of a crane.

The maximum load capacity of a crane W_{max} to prevent tipping is as follows:

$$W_{max} = \frac{W_{cr}(a + b)}{r_{min} - b} \qquad (6.1)$$

where W_{cr} is the weight of the crane, including the jib, and the distances a, b, and r_{min} are shown in Figure 6.1.

To incorporate a safety factor for a crane standing on firm level ground the rating, the safe working load, should not exceed a certain percentage[9-11] of the load which produces tipping. USAS B 30.5:1968 states that the working load for a crawler mounted crane should not exceed 75% of the tipping load without outriggers or 85% of the tipping load if the outriggers of the crawler crane are fully extended. BS 1757:1964 states that for crawler mounted cranes the working loads should not exceed 67% of the tipping load for working loads up to about 8 Tonnes (8000 kg) and 75% of the tipping load for working loads over 12 Tonnes (12000 kg). For loads from 8 to 12 Tonnes the rating should vary from 67 to 75% of the tipping load. The total weight of the lifting hook, slings, spreader beams, and the load carried must not exceed the safe working load.

6.2 TYPES OF LIFTING EQUIPMENT

The types of machines used on site for lifting purposes are as follows:

1. A crawler mounted crane
2. A mobile crane
3. A tower crane
4. A forklift truck
5. A derrick

6.3 CRAWLER MOUNTED CRANES

On most cranes, except for tower cranes, the hook radius is altered by pivoting the jib (referred to as luffing) in a vertical plane about its lower end where the pivot is mounted at the base. The lifting radius and height of a crane can be increased by attaching a small flyjib to the end of the main jib. Figure 6.2 shows the ranges of height and radius for various selected jib lengths of a particular crane. The extended height and radius is also shown when a fly jib is used. Table 6.1 shows typical radius-height-capacity values for a particular crawler mounted crane. When a flyjib extension is used, the capacity is reduced for the shorter main jib lengths, but this reduction is less significant in the longer main jib lengths. Special care is necessary when raising long main jib–flyjib combinations from ground level or when working at low angles, particularly when the ground conditions are soft or uneven.

TABLE 6.1   Main Jib Capacities of a Particular Crane

	Jib length (m)				
Radius (m)	9.0	18.0	24.0	30.0	36.0
3.10	35.5				
4.50	23.7				
6.00	15.5	15.2			
9.00	9.0	8.7	8.5	8.3	8.1
18.00		3.4	3.2	2.9	2.7
24.00			2.0	1.8	1.5
30.00				1.1	0.9

Typical approximate main jib capacities
(kg × 10³ or Tonnes)

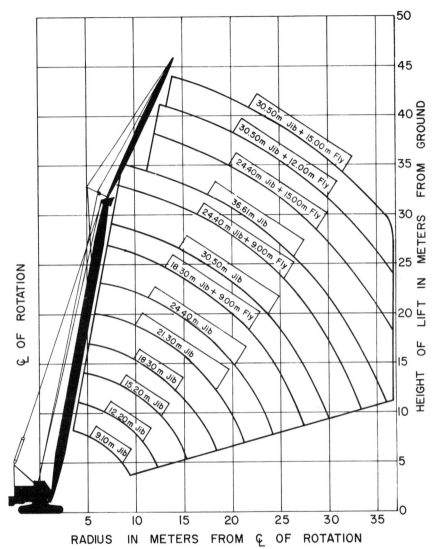

Figure 6.2 Crawler mounted crane, height versus radius.

There are several advantages in using crawler mounted cranes on site. The crawler tracks are robust, suitable for rugged conditions, exert a low pressure on the ground, and are therefore suitable for operating on soft muddy ground likely to be encountered on construction sites. The tracks contribute to a fairly stable mounting with good maneuverability and are capable of climbing steep slopes up to about 1

TABLE 6.2   Crane Ratings Versus Jib
Lengths for a Crawler Mounted Crane

Crane ratings ($\times 10^3$ kg)	Jib lengths (m)	
	Basic	Maximum
6	9	18
18	9	36
30	10	35
55	12	49
110	15	61

in 3. Compared with other types of cranes, crawler mounted cranes
have the lowest initial capital cost.

The assembly of the crawler mounted crane takes several hours, and
the jib shortening or lengthening operation, in which one or more jib
sections are removed or added, is also time consuming. A long space on
site is required for assembly or modification to the jib. Transportation
from one construction site to another may involve several vehicles, and
the travel speed on the site itself is low, usually about 3 km/hr.

Table 6.2 shows a typical range of ratings and jib length ranges for
crawler cranes. The values give an overall approximate indication of
what capacities are available including, in some cases, an allowance
for a flyjib extension. Further details are necessary when selecting a
crane for a specific site, and the crane manufacturer's specifications
should be consulted and compared.

6.4   TRUCK MOUNTED MOBILE CRANES

Recent advances in the design of truck mounted mobile cranes have
made this type of crane a more likely choice of crane on a construction
site. The telescopic boom types permit immediate changes in the capacity, height, and radius. Traveling from one site to another is simplified because the truck unit is capable of fast traveling speeds at
about 60 km/hr and, on arrival, is almost immediately available for
use. It can be used for various general lifting activities at different
locations on site in any one day.

The truck mounted mobile crane has, however, a higher capital cost
than the equivalent crawler unit. The heavy axle loading on soft

ground limits its use and it cannot travel easily over very rough terrain. Half tracks can be fitted to some types to improve performance on soft ground. Short movements at a particular location can be slow when there are separate power units for the truck and crane, and the outriggers, if used, also take time to readjust.

The height versus radius of a typical truck mounted mobile crane is shown in Figure 6.3.

Table 6.3 shows the ranges of capacities and radii available for truck mounted mobile cranes with telescopic booms and with the standard jib sections in fixed lengths. The capacity obviously alters for a particular crane depending on radius and jib length. The typical capacity ranges

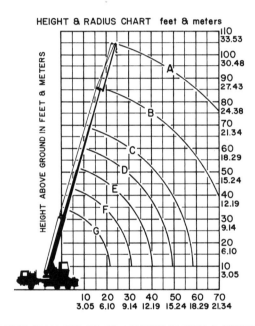

HEIGHT & RADIUS CHART   feet & meters

DISTANCE FROM CENTER OF ROTATION IN FEET & METERS

(A)   80' (24.38 m) Boom + 20' (6.10 m) Fly Jib
(B)   80' (24.38 m) Boom
(C)   62' (18.89 m) Boom
(D)   53' (16.15 m) Boom
(E)   44' (13.41 m) Boom
(F)   35' (10.67 m) Boom
(G)   26' (7.93 m) Boom

Figure 6.3 A truck mounted mobile crane, height versus radius.

TABLE 6.3   Capacity Versus Jib Length and Radius for Truck Mounted Mobile Cranes

Crane capacity ($\times 10^3$ kg)	Jib length (m)		Radius (m)	
	From	To	From	To
12	8	25	3	15
20	9	25	3	18
30	10	30	3	24
52	12	35	3	24
108	13	59	3	50
Telescopic booms				
55	8	54	4	36
125	12	72	4	50
230	12	95	4	60
Standard jib sections				

TABLE 6.4   Capacities of a 125 Tonne Truck Mounted Mobile Crane

Radius (m)	Jib length (m)				
	12.0	24.0	42.0	60.0	72.0
3.5	143				
6.0	117	95			
9.0	69	69	54	30	
12.0	44	44	44	28	17
18.0		24	24	20	15
24.0		16	15	14	12
30.0			10	9	7
50.0				2	1
Capacities (Tonnes or $\times 10^3$ kg)					

for a 125 Tonne truck mounted mobile crane with a standard jib section are shown in Table 6.4.

6.5   SELF-PROPELLED MOBILE CRANES

The self-propelled mobile crane generally has a smaller range of capacities and characteristics compared with the truck mounted mobile

crane, but it has a lower capital cost for a given rating. In comparison
with the crawler mounted crane it has a higher traveling speed on site,
about 25 km/hr, and a higher capital cost. It cannot travel easily over
rough terrain, and a truck trailer is required for transportation from
one site to another.

6.6 TOWER CRANES

Some tower cranes have luffing jibs that pivot at the top of the tower to
alter the hook radius so that the tower can be erected close to the
structure under construction. Generally, however, tower cranes do not

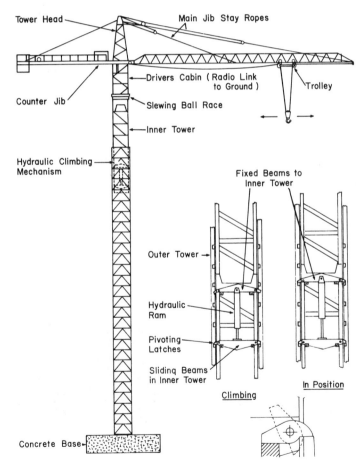

Figure 6.4 A Tower crane.

have luffing jibs, but have a permanently horizontal jib from which a trolley moves in and out from the tower along the jib to alter the hook radius. Refer to Figure 6.4. This type of jib enables the tower to be erected very closely to the structure under construction and may often be erected within the structure in a suitable elevator shaft or other similar position.

The crane may consist of an inner and outer tower and is then self-climbing. Both towers are usually latticed members. The first outer section is fixed to the base, and the first and second inner sections are fixed into the outer frame. The lower inside section contains hydraulic

Plate 6.1   A crawler mounted crane. Courtesy T. Smith & Sons Ltd., Leeds, W. Yorks, England.

jacks and the climbing mechanism. Some types do not contain a self-climbing mechanism.

The slewing ring, slewing platform, and tower head are then assembled. The main jib is lifted and bolted to the tower head with the jib supported off the ground. Then the counterjib is lifted into place and supported by a stayrope and the jibs are raised to the horizontal.

At this point the crane is now self-erecting and must be tested to an overload of 25%. The crane itself lifts the 6.4 m outer sections into place, and afterward the hydraulic rams then lift the inner tower upward, complete with the jib, while the pivoting latches on the rams lock onto the crossmembers of the outer tower or onto collars attached to the floors of the building.

Plate 6.2   A crawler mounted crane. Courtesy T. Smith & Sons Ltd., Leeds, W. Yorks, England.

Plate 6.3 A truck mounted mobile crane. Courtesy T. Smith & Sons Ltd., Leeds, W. Yorks, England.

More outer sections may be added to increase the height of the crane to about 50 m, above which they must be tied and can be then extended to about 150 m. Wind effects must be carefully considered for a tower crane. The tie frame is specially constructed to resist dangerous movements of the tower from wind blowing from any direction.

To provide greater mobility tower cranes are sometimes mounted on rails, trucks, or a crawler chassis. A portable remote control unit enables the crane to be operated when the view is obstructed from the operator's cabin.

A tower crane with a 49 m long jib and a counterjib 25 m long can lift a load of 15×10^3 kg to a radius of 14 m; the load capacity then decreases to about 3×10^3 kg at a 47 m radius. A smaller sized crane with a 24 m long jib and a counterjib 13 m long can lift a load of 8×10^3 kg to a radius of 12 m, decreasing to about 3×10^3 kg at a 24 m radius.

Plate 6.4 A truck mounted mobile crane. Courtesy T. Smith & Sons Ltd., Leeds, W. Yorks, England.

Plate 6.5 A self-propelled mobile crane. Courtesy Coles Cranes Ltd., Uxbridge, Middlesex, England.

Plate 6.6 A tower crane. Courtesy Koehring Crane and Excavator Group, Milwaukee, Wisc.

Plate 6.7   A forklift. Courtesy Winget Ltd., Rochester, Kent, England.

6.7 FORKLIFTS

A forklift attachment on a tractor enables loads up to about 3000 kg to be lifted to a height of 5.5 m. It can handle many different types of materials on construction sites with speed and economy. In top gear the forklift can travel at a speed of 25 km/hr. In addition to the double mast forklift a range of attachments can make this type of machine an extremely useful lifting and transporting machine on site. A load extender can be fitted, but this reduces the load capacity from 2000 kg, at a load center from the mast of 500 mm, to a load capacity of 1250 kg, at a 1.2 m eccentric load center, for a typical machine.

6.8 DERRICKS

A derrick is a lifting machine that possesses a vertical mast or A frame secured by inclined stiff legs or guy ropes. The boom pivots about the

Plate 6.8 A derrick crane. Courtesy Butter Cranes Ltd., Hillingdon, Middlesex, England.

bottom of the mast, and its inclination is adjusted by wire ropes attached from the top of the mast to the top of the boom. With the advent of tower cranes, derricks are now virtually obsolete except when very heavy loads must be lifted or when the derrick remains permanently in one place. Occasionally there are conditions on a construction site which still make selection of a rail mounted derrick appropriate.

CHAPTER 7

EXCAVATING EQUIPMENT

7.1 TYPES OF EXCAVATING EQUIPMENT

The types of equipment used on a civil engineering site for excavating material are as follows:

1. Face shovel (power shovel)
2. Dragline
3. Grab
4. Hydraulic excavator (backhoe, backactor, drag shovel)
5. Excavator loader
6. Front end loader (tractor shovel)
7. Trenching machine
8. Ripper

7.2 FACE SHOVEL (POWER SHOVEL)

A typical face shovel is shown in Figure 7.1. The hydraulic or cable activated unit is attached to a crawler crane unit. The machine unit can be converted from a crane to a face shovel, hydraulic excavator, dragline, or grab, although the large mining machines are specifically designed for a particular job with no provision for interchangeable attachments. The shovel digs mainly above the base level of the machine by pushing the bucket away from the machine and through the bank of the material to be excavated. The bucket is then raised and the material is generally loaded into trucks through a back gate in the bucket. Ripper and skimmer attachments are also available instead of the shovel bucket.

The cycle time for the face shovel is considered to be fairly fast when

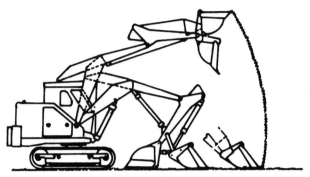

Figure 7.1   Face shovel.

an easy sight of the excavating and dumping positions is possible. Cycle times[12] vary according to the type of machine, the nature and unit weight of the excavated material, the angle of swing, the bucket size, and the height of the excavation.

The production in cubic meters per hour based on in situ volumes is given for various sizes of buckets in Table 7.1 for a 90° swing, an optimum digging depth, and no delays. There should be no delays if the bucket size is matched with the available truck capacity.

The optimum digging height is shown in Table 7.2 for a $\frac{3}{4}$ and $1\frac{1}{2}$ m^3 bucket size.

Correction factors for the actual height of excavation, the angle of swing, and carry factors for a particular machine should be taken from the manufacturer's handbook.

The disadvantages and advantages of self-propelled, truck mounted,

TABLE 7.1   Face Shovel Production

| | Bucket size (m³) | | |
Material	$\frac{3}{4}$	$1\frac{1}{2}$	3
Sand and gravel	150	250	420
Earth	130	230	390
Clay, hard tough	110	200	350
Well blasted rock	95	175	315
Clay, sticky wet	75	125	265
Poorly blasted rock	60	110	230
(see reference 12)	Production (m³/hr)		

TABLE 7.2 Optimum Heights for Face Shovel

Bucket size (m³)	Optimum height (m)		
	Sand gravel	Common earth	Hard rock or sticky clay
$\frac{3}{4}$	1.8	2.4	2.7
$1\frac{1}{2}$	2.4	3.1	3.7

and crawler mounted general excavating machines are similar to those described for cranes in Chapter 6.

Over the last few years the excavation of a face on a construction site has veen carried out increasingly by a front end loader machine (see Section 7.7) except for the excavation of very hard rock and for high faces, such as in some quarries.

7.3 DRAGLINE

The dragline has the greatest digging reach, digging depth, and disposal range compared with all the attachments that can be fitted on a universal machine, but it has less digging power than an hydraulic excavator. A typical machine is shown in Figure 7.2. Bucket capacities

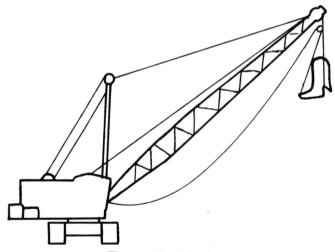

Figure 7.2 A dragline.

on civil engineering sites usually vary between 0.5 and 2 m³, although exceptionally large buckets are used on open pit coal mining, tar sands, or quarrying sites with capacities ranging from 6 to 65 m³. The large machines are sometimes of the walking type.

The dragline is an excavator with a crane boom and a crawler carriage with an A frame and fairlead attachment. It is used for jobs such as trimming embankments, excavating loose material in sand or gravel pits, and dredging operations. The excavation can be performed many meters below the base level of the machine, and the bucket can also be cast to a spot a considerable horizontal distance beyond the end of the boom, although this is not recommended in certain circumstances. The cycle time for the dragline tends to be slower than the shovel, and the rate of excavation deteriorates when the angle of swing increases.

Greater stability can be achieved by the use of additional counterweights and wider or longer tracks. Even with a banksman, bucket placing can be difficult and dumping may not be precise. The dragline usually has a longer life than the shovel and, therefore, has a lower hire rate.

Table 7.3 shows the working ranges of a dragline for various boom lengths. The operating radius is the horizontal distance from the center of the machine tracks to the end of the boom, and the clearance

TABLE 7.3   Working Ranges for Various Boom Lengths
for a Dragline

Length of boom (m)	12	15	18	21	24	27
Operating radius (m)	11	13	16	18	21	23
Clearance height (m)	10	12	14	16	18	20

TABLE 7.4   Bucket Size Versus Gross Load Versus
Radius for a Dragline

Bucket size (cu/m)	Bucket weight (kg)	Gross load (kg)	Normal maximum dumping radius (m)
1.0	1150	2860	23
1.2	1220	3760	21
1.6	1630	4350	17
2.0	1900	5700	15

TABLE 7.5 Dragline Production (m³)

Soil	Bucket size (m³)			
	1.0	1.2	1.6	2.0
Sandy clay, loam	150	165	200	230
Sand and gravel	140	160	195	225
Earth	135	145	175	200
Clay, hard	100	120	150	175
Clay, wet and sticky	70	85	110	135

height is measured from the top of the boom to the machine base level. Note that the bottom of the bucket is several meters below the top of the boom. Table 7.4 shows the gross load for various bucket sizes.

An hourly production for draglines is given in Table 7.5 for various bucket sizes. These values are only approximate and should be corrected for the difference between the depth of excavation and the optimum depth, and the angle of swing.

7.4 GRAB (CLAMSHELL)

The capacity given in Table 7.6 includes the weight of the grab. Compared with cranes, the capacity as a percentage of the tipping load is reduced for grabs and draglines.

Almost the same boom attachments are used for the grab as for the

TABLE 7.6 Grab—Ranges and Capacities

Material	Grab size (m³)	Empty weight (kg)	Loaded weight (kg)	Working radius (m)
Ex	1.0/0.8	1800	3500	17
S	1.5/1.25	1600	3700	17
Ex	1.5/1.25	2800	5300	13
S	2.0/1.5	1900	5100	13
Ex	2.0/1.5	3600	6800	11
C	2.6/2.1	1800	3900	16
S	2.6/2.1	2600	6800	11
C	4.0/3.1	2800	6000	11

Ex = excavator/excavated material; S = sand type; C = coal type; counterweight 11,200 kg.

Figure 7.3 A grab bucket.

dragline. The grab usually consists of a hinged bucket in two shells or halves, see Figure 7.3, or may contain two, three, or several jaws. When the drop weight grab is attached to the universal crane unit, it develops its digging power by the dynamic force of the open jaws of the bucket suddenly dropping to the spot to be excavated.

The grab can excavate materials out of stockpiles and excavate in subaqueous conditions, as well as being able to discharge its load at a considerable height above the base level of the machine. It excavates areas more accurately than the dragline and can dig deep and narrow holes. The grab cycle time is slower than the dragline because of the time taken to close the bucket and because the load is always carried at the extreme end of the boom.

The hydraulically operated grab has two opposing rams, one to each shell, which drive the jaws. It is more effective than the drop weight grab over shorter operating radii. Figure 7.4 shows a diagrammatic sketch of an hydraulic grab, and Table 7.7 shows the work-

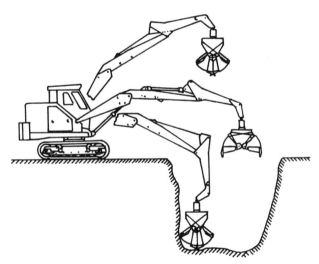

Figure 7.4 An hydraulically operated grab.

TABLE 7.7   Working Ranges of a Typical Hydraulic Grab

	Working dimensions (m)		
Effective height above base level	3.3	4.5	1.3
Effective depth below base level	4.7	3.5	5.2
Operating radius	7.5	7.5	7.5

ing ranges of a typical hydraulic grab; the variations in the distances arising from the variable boom and ram positions.

7.5   HYDRAULIC EXCAVATOR

The hydraulic excavator (Figure 7.5) is frequently referred to as a drag shovel, backhoe, or backactor. The features of the hydraulic excavator are similar to the face shovel except that it digs mainly below the base level of the machine by dragging the bucket toward the machine. It is, therefore, most suitable for trench excavation, and it can excavate to a considerable depth below the machine. Accurate placing and sighting of the bucket are possible with the machine. The cycle time is slightly slower than the face shovel because of the extra time to discharge from

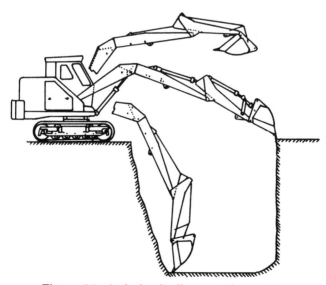

Figure 7.5   An hydraulically operated excavator.

the bucket. The hydraulic excavator production is not as high as trenching machines, but it can operate in steeper, more rugged conditions and in harder material.

The production rate depends on the struck, or heaped, capacity of the bucket, the percentage of voids in the excavated material, and the cycle time. The cycle time varies according to the degree of difficulty encountered in excavating the material, the swing angle, the depth of the excavation, the height of loading, and the nature and type of soil. Carry factors are shown in Table 7.8 for various materials.

Typical maximum loading heights range between 3.0 and 8.0 m, and maximum digging depths range between 3.5 and 10.0 m for various machines. Maximum operating radii range between 7.0 and 14.0 m.

The production[12] for various bucket sizes in various degrees of digging difficulties is shown in Table 7.9.

TABLE 7.8   Carry Factors

Material	Carry factor (%)
Moist loam or sandy clay	100–110
Sand and gravel	90–100
Hard tough clay	75– 85
Rock, well blasted	60– 75
Rock, poorly blasted	40– 50

TABLE 7.9   Production—Hydraulic Excavator

Estimated cycle		Bucket size					
Time (sec)	No./min	0.5	1.0	1.5	2.0	2.5	Zone
15	4	120	240	360			A
20	3	90	180	270	360	450	
30	2	60	120	180	240	300	B
40	1½			135	180	225	C

Production (m³) at 100% efficiency.

A = easy digging, depth ⧴ 30% maximum depth, 30° swing; B = average digging, depth ⧴ 50% maximum depth, 90° swing; C = tough digging, depth > 75% maximum depth, swing > 120°.

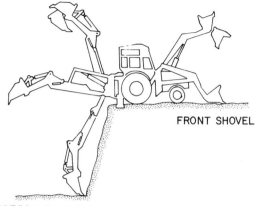

FRONT SHOVEL

HYDRAULIC EXCAVATOR

Figure 7.6   An excavator-loader.

7.6   EXCAVATOR-LOADER

The smaller multipurpose machine, shown in Figure 7.6, with a loading shovel at the front and an hydraulic excavator, face shovel, or hydraulic powered rotating grab at the back of the machine is a useful machine because of its versatility in performing the many different jobs usually encountered on a construction site. Hydraulic hammers and other specialized equipment can also be mounted on some machines. Typical ranges of various machines are given in Table 7.10.

7.7   FRONT END LOADER

Figure 7.7 shows a typical front end loader or tractor shovel in the smaller bucket capacity range. Over the last decade, because of its mobility, maneuverability, and versatility, the front end loader has

TABLE 7.10   Working Dimensions—Excavator-Loader

Shovel capacity (m³)	0.7	0.7	1.0	1.25
Loading height (m)	3.2	3.4	3.4	4.2
Digging depth (m)	3.4	3.7	4.1	4.7
Reach (m)	4.9	5.3	5.5	6.2

Figure 7.7 A front end loader.

substantially replaced the face shovel in the operation of removing material from the face of an excavation.

Front end loaders can be mounted on a chassis with either tracks or rubber tires. They are particularly useful for load and transport operations by themselves over short distances or with attendant trucks over longer distances.

The front end loader has a good lift speed and bucket reaching height and the rubber tired type can operate at about 25 to 40 km/hr in forward and reverse gears. Some types of loader are articulated within an angle of 70°, that is, 35° to the right and left.

For a 4.5 m³ loader the forward and reverse speed in first gear is about 7 km/hr, and the speed in the second forward and reverse gears is about 13 km/hr, the third gear being used primarily for transportation.

The smaller types have bucket capacities within the range of 1.3 to 3.0 m³ while the larger range bucket capacities vary from 3 to 7.5 m³. For estimating purposes a carry factor of about 0.9 should be considered depending, of course, on the type and nature of the material being excavated.

In good circumstances a machine with a 1.6 m³ bucket can load about 200 m³/hr, whereas a machine with a 7.5 m³ bucket can load about 1000 m³/hr.

7.8 TRENCHING MACHINE

Increasing use is made of the special purpose continuous chain trenching machine, shown in Figures 7.8 and 7.9. It is most suitable for long, shallow, narrow trenches where the excavated material is not too

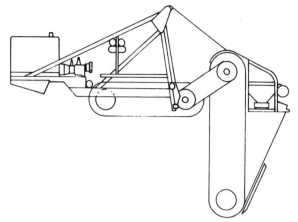

Figure 7.8   A Trenching machine.

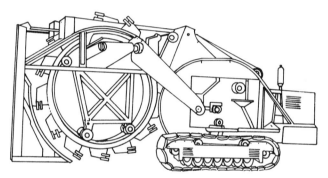

Figure 7.9   A wheel type trenching machine.

dense and tough. The machine excavates continuously by means of buckets attached to a chain which rotates and is usually shaped like a ladder. A hinged boom is attached to a specially equipped tractor unit to carry the chain mechanism, and the spoil is cast aside at the side of the trench by means of a conveyor belt.

Although the buckets have a tendency to twist due to the lack of torsional resistance of the chain, the machine can achieve a relatively fast digging speed. The depth, and particularly the width, can be controlled accurately, and therefore finishing by hand is reduced, and unnecessary excavation is almost eliminated.

Typical data for a trenching machine are given in Table 7.11. The digging speeds can vary appreciably depending on the toughness of the soil and the depth of the trench.

TABLE 7.11   Trenching Machine Data

Maximum trench depth	3 m
Digging speed	0 to 10 m/min
Travel speed, top gear	4.6 km/hr
Trench width	0.48–0.61 m

Another type of trenching machine, shown in Figure 7.9, consists of a hinged steel frame carrying a wheel suspended from the special tractor unit. Buckets are attached to the periphery of the wheel.

Stability can be a problem with trenching machines because of the cantilever loads of the buckets and conveyor belt. The tractor engine is therefore cantilevered at the other end of the machine as a counterbalance. The wheel type machine may be widened with tapered cutting teeth to form an excavated side slope to the trench if this is required.

7.9   RIPPER ATTACHMENTS

A ripper attachment is shown in Figure 7.10. It is generally mounted on and pivots about the rear end of a crawler tractor. Up to five ripper shanks may be used, and the maximum penetration can be up to about 400 mm depending on the type of ripper attachment and material.

The ripping power of the attachment depends on the properties of the material being ripped and the power and weight of the tractor, as well as the penetrating force of the ripper attachment.

The ripper attachment is particularly suitable for breaking up shales and slates and is often used on asphalt, hard clay, and frozen

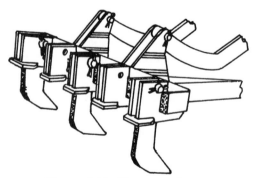

Figure 7.10   Ripper attachment.

ground. The ripping operation frequently puts a heavy work load on the tractor.

7.10 WORKED EXAMPLES

Example 7.1

A face shovel with a $1\frac{1}{2}$ m³ bucket is to load well blasted rock into 25 m³ haul trucks. If the shovel is working at the optimum height, determine the weekly production of the shovel and the number of trucks required if the cycle time for the trucks is 30 min, including an efficiency allowance. Assume a 45 hr week. The following data, which can be found in the machine manufacturers' handbooks, are required:

Good job conditions efficiency   = 75%
Swing 60° swing factor   = 1.16%
Carry factor for well blasted rock = 75%

Plate 7.1 A face shovel (background). Courtesy of Ruston-Bucyrus Ltd., Lincoln, England.

Plate 7.2   A dragline. Courtesy T. Smith & Sons Ltd., Leeds, W. Yorks, England.

Answer 7.1

Uncorrected hourly production = 175 m³ (see Table 7.1)
Reduction factor   = 0.75 × 0.75 × 1.16
                                           = 0.65
Average hourly production   = 175 × 0.65
                                           = 114 m³ in situ volume
Average weekly production   = 114 × 45
                                           = 5130 m³

During the loading operation the shovel is assumed to be 100% efficient to ensure that it is not idly waiting for trucks.

No. of trucks required per hour   = $\dfrac{114/0.75}{25}$

                                           = 6.1 trucks per hour

Plate 7.3 A drop clamshell (grab). Courtesy T. Smith & Sons Ltd., Leeds, W. Yorks, England.

The cycle time for the trucks is 30 min, including an efficiency factor.

$$\therefore \text{ No. of trucks required} = 3$$

Example 7.2

Determine the weekly production of an articulated wheel loader equipped with a rated 5 m³ bucket that loads material, travels 96 m in second gear (assume 4 m/sec), and then loads the material into trucks.

Plate 7.4 An hydraulic clamshell. Courtesy Drott (J.I. Case/Tenneco Inc.) Racine, Wisc.

Plate 7.5 An hydraulic excavator. Courtesy Drott (J.I. Case/Tenneco Inc.) Racine, Wisc.

Plate 7.6 An excavator-loader. Courtesy JCB (Canada) Excavators Ltd., Burlington, Ont., Canada.

Plate 7.7 A front end loader. Courtesy Volvo Canada Ltd., Bramalea, Ont., Canada.

Plate 7.8 A front end loader. Courtesy International Harvester, London, England.

Plate 7.9   A track front end loader. Courtesy Caterpillar Tractor Co., Peoria, Ill.

A carry factor of 0.85, a job efficiency of 0.75, and a 40 hr week may be assumed.

Answer 7.2

To simplify wheel loader production calculations the cycle time is assumed to consist of a basic cycle time plus the haul and return journeys. The basic cycle time includes loading, dumping, and maneuvering, which for an articulated loader can be assumed to be 0.4 min, but with corrections for certain site conditions.

Basic cycle time		= 0.40 min
Haul time	= 96/4 × 60	= 0.40 min
Return time		= 0.40 min
Correction to basic cycle time		= 0.00 min
Cycle time		= 1.20 min
No. of cycles per hour	= 60/1.20	= 50
(uncorrected)		
Loose load	= 5 × 0.85	= 4.25 m²
Production/week	= 40 × 50 × 4.25 × 0.75	= 6375 m³

Plate 7.10 A trenching machine (pipeline ditcher). Courtesy Barber-Greene Co., Aurora, Ill.

Plate 7.11 A trenching machine (pipeline ditcher). Courtesy Barber-Greene Co., Aurora, Ill.

Plate 7.12   A trenching machine (pipeline ditcher). Courtesy Barber-Greene Co., Aurora, Ill.

Plate 7.13   A ripper attachment. Courtesy Terex, G. M., Hudson, Ohio.

Example 7.3

A job on the critical path requires the removal of 10,000 m³ of soil in 10 days. Two hydraulic excavators are available, one equipped with a 1.0 m³ bucket and the other equipped with a 1.5 m³ bucket. If average digging conditions are assumed, in which the depth does not exceed 50% of the maximum digging depth, and a 90° swing is predicted, determine which machine should be used for the job. A coefficient of bulking of 0.25 and a 10 hr day should be assumed.

Answer 7.3

In situ volume	= 10,000 m³
Loose volume	= 12,500 m³

$$\text{Uncorrected volume to be removed per hour} = \frac{12,500}{10 \times 10}$$
$$= 125 \text{ m}^3$$

By referring to Table 7.9 it can be seen that the hydraulic excavator with the 1.5 m³ bucket should be selected, as this machine can remove $180 \times 0.83 = 149$ m³/hr in the prescribed conditions.

Note: It can be seen in Examples 7.1 to 7.3 that the solutions require a considerable amount of data which can normally be obtained for particular machines from sources such as the manufacturers' handbooks.

CHAPTER 8

EXCAVATING AND TRANSPORTING EQUIPMENT

8.1 TYPES OF EXCAVATING AND TRANSPORTING EQUIPMENT

The types of equipment that excavate and haul materials are as follows:

Loaders

Scrapers

Bulldozers

Graders

Many parameters influence the choice of equipment for bulk earth moving operations. As each individual site varies, the contractor must decide which type of machine is most suitable to excavate and haul the material at the most economical cost per cubic meter.

For short distances up to about 100 m a bulldozer is likely to be the most suitable if the material is fairly loose or has been broken up. The bulldozer is particularly useful when the material has to be hauled over a rough or muddy surface for a short distance and spread evenly.

If the material has to be hauled up to about 200 m and is soft, or is to be excavated from a stockpile or dumped into a truck or onto a stockpile, a loader is suitable. For shorter travel distances where the haul route is rough or muddy and the loading operation requires a lot of power a track type loader should be selected. For longer distances where the stockpile is fairly easy to load and where the haul road is reasonable a wheel type is the most economical. Loaders and bulldozers are described in Sections 7.7 and 8.4, respectively.

Trucks are used for long haul distances several kilometers or longer or where the material is bulky such as broken lumps of hard rock. If the material has to be transported on highways, trucks are necessary for even shorter distances. Attendant plant is required to load and perhaps spread the material at each end of the haul route. Shovels, loaders, and draglines may be used for the loading operation, and bulldozers or graders may be used for the spreading operation. The choice of the attendant plant depends on the type of material and its disposition. Belt conveyors, monorails, and cableways are alternative methods worth considering for long haul distances. Further details of hauling equipment can be found in Sections 8.6 to 8.10.

For medium haul distances up to about 2 km, and where the haul route is within or near the contract limits, scrapers are generally employed because they are the most economical excavate-and-haul equipment for large earth moving operations except where the route is sited on or across the public highway or where the soil properties prevent their use. Various types of scrapers are shown in Figure 8.1.

There are several types of scrapers such as track type tractor scrapers, single or twin engined wheel-type tractor scrapers with two or three axles, and elevating scrapers. A scraper is normally loaded as it moves forward by lowering the hydraulically operated cutting edge of the scraper bowl, and the front apron is raised, either simultaneously or separately controlled, to provide an opening through which the scraped material is forced into the bowl. When the loading operation is complete, the cutting edge is raised and the apron lowered. In the dumping operation the cutting edge is lowered to the required height for the depth of fill required, and the apron is raised. An ejector blade forces the material out and helps control the dumping operation. The ejector blade is also used in the loading operation to stack the scraped material in the bowl.

The elevating scraper is equipped with an elevator frame with about 15 flights which scoop up the scraped material and deposit it on top of the load within the scraper bowl. This action requires less power than a conventional scraper and therefore usually obviates the use of a pusher tractor. The flights also break down the scraped material, which results in an increase in the material hauled per cycle and a more effective spreading operation.

8.2 THE SELECTION OF SCRAPERS

Track type scrapers are less frequently used today. A track type might be considered, if available, when the material must be scraped, hauled,

Standard 2 Axle Scraper   Standard 3 Axle Scraper

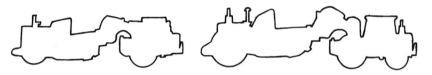

Twin Engined 2 Axle Scraper   Twin Engined 3 Axle Scraper

Elevating Scraper   Tractor Drawn Scraper

Push-Pull Scraper

Figure 8.1 Types of scrapers.

and spread over short haul distances of about 100 to 500 m, and when a rubber tired scraper would not be able to maintain its highest operational speed for very long. A track type scraper might also be preferred when traction problems arise due to very poor ground conditions.

When top speed hauling is required, the standard single engined scraper is considered to be the most economical machine over comparatively medium and long hauls where well maintained haul roads provide low rolling resistance, average traction conditions, and where zero or moderate adverse gradients occur.

The single engined scraper is the most versatile scraper in all soil types. The Caterpillar Tractor Co. suggests that the single engined scraper can work in classes of soil including clay, silt, sand, gravel, and earth-rock mixtures. Standard scrapers can only marginally work in rock that has been ripped or blasted. Other types of rubber tired scrapers are less versatile, the least versatile being the push-pull type of scraper, where slippage can become a problem and tire costs increase substantially.

Twin engined scrapers, where front and rear engines provide power for all the wheels, can operate in difficult ground conditions such as soft muddy areas and uncompacted fill areas. This type of scraper can climb gradients steeper than 1 in 3 on moderately firm ground. Twin engined scrapers can operate in adverse weather conditions because of their better traction and retarding ability. Single and twin engined scrapers usually require a crawler tractor to push load them to obtain an optimum production and cost benefit.

The most profitable application of elevating scrapers occurs when the haul distances are less than 1 km and when the haul road is relatively level and has a low rolling resistance. Elevating scrapers work well alone, where a large pushed scraper fleet is not economical, because of the planned rate of excavation, or where the amount of excavation is not large.

Push-pull scrapers consist of two twin-engined scrapers attached together with a special hook arrangement. During the loading part of the cycle one scraper either pushes or pulls the other scraper, thus creating more power without the aid of a crawler tractor pushing.

Scraper bowls vary in heaped capacity from about 6 to 40 m³. Unless scrapers are available from another site, where the major earthworks operations have been completed, the size of the scrapers is determined by the size of the earth moving project. Usually smaller scrapers with heaped bowl capacities of about 15 m³ are used on projects where about 1 million cubic meters of soil is to be transported. On larger projects, where 4 million cubic meters or more of soil is to be transported, scrapers with heaped capacities of 30 m³ or more are frequently used. The larger scrapers increase production because larger loads are carried in nearly the same cycle time. Larger scrapers lower the unit production costs because operator costs remain the same and there are only small increases in owning and operating costs. However, larger scrapers must fully utilize their productive capacity for optimum economy. They also require more room for maneuvering and more pushing power.

Two axle scrapers possess better traction and maneuverability characteristics and are preferable to three axle scrapers except for long haul distances.

8.3 SCRAPER CAPACITIES AND PRODUCTION

Table 8.1 shows the ranges of the capacities of scrapers. The push-pull and elevating types are usually self-loading. The struck capacity is the volume of the bowl with the excavated material level with the top of the bowl. The heaped capacity is the volume with the top of the excavated material heaped above the bowl and sloping down to the edges of the top of the bowl. The heaped capacity is usually considered for estimating purposes.

To estimate the hourly production of a scraper, reference should be made to Sections 5.2 to 5.4. The time taken to complete one cycle is the sum of the load, haul, dump, and return times. The load, dump, maneuver, acceleration, and deceleration times are usually considered as fixed times and depend on the type and capacity of the scraper and whether it is self-loading or being pushed. The haul and return times are considered as variable times because the times directly depend on the speed of the scraper and on the length of the haul and return journeys.

By calculating the rimpull required, the speed of the scraper can be determined using a graph similar to the speed versus rimpull graph shown in Section 5.7. Graphs for individual machines may be obtained from the plant manufacturers or suppliers. The total weight of the load carried by a scraper should be checked so that it does not exceed the rated load of the scraper. The rated load is the weight capacity of the scraper, which may occupy less volume than the heaped capacity of the scraper depending on the unit weight of the transported material.

Scraper operations inherently cause dust problems with many types of soil in dry weather conditions. In areas where other machines are operating or site personnel are on foot, dust can constitute a great

TABLE 8.1   Range of Scraper Capacities

Type	Capacities (m³)	
	Approximate struck	Approximate heaped
2 No. twin engined	50	68
	37	50
Push-pull	27	37
	21	30
3 axle		
Twin engined	31	41
Single engined	31	41
	25	34
	16	23
2 axle		
Twin engined	25	34
	18	24
	14	19
	11	15
Single engined	25	34
	21	29
	18	24
	16	22
	11	15
Elevating		27
		17
		8
Crawler tractor	18	23
scraper	13	17
	9	11
	5	6

danger due to poor visibility. In this situation water spraying machines should be used to alleviate the dust problem.

Worked examples calculating the production of scrapers are given in Section 8.11.

8.4   BULLDOZER AND ANGLEDOZER

The bulldozer (Figure 8.2) is used primarily in spreading and leveling operations for distances up to about 100 m. As the tractor pushes for-

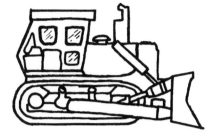

Figure 8.2 Track type bulldozer.

ward, the bulldozer blade digs into the ground and gradually accumulates a wedge shaped load over a distance of about 10 to 20 m. Two types of blade configurations are usually used for dozing operations: the straight blade and the universal blade, which is curved in plan and elevation.

The bulldozer is a very rugged and versatile machine which can excavate most types of material. Even hard rock, which has been broken by ripping or blasting operations, can be fairly easily moved by a bulldozer. The production of this type of machine depends on many variables such as dozing distance, operator and job efficiency, size of machine, type of material, machine cutting distance, and type of blade.

Table 8.2 shows an approximate loose measure production rate for track type bulldozers with average site conditions and 100% efficiency. The rates for a specific machine should be altered accordingly with correction factors from the manufacturer's handbook for conditions that are not average.

For less rugged jobs rubber tired tractors (Figure 8.3) are generally preferred to track mounted machines because of their increased speed and maneuverability.

Normally the blade of a bulldozer lies in a line perpendicular to the direction of movement. The blade of an angledozer is inclined to this line and is very useful for backfilling trenches and surface stripping

TABLE 8.2 Approximate Hourly Production of a Track Type Bulldozer

Average dozing distance (m)	50	75	100	150
Approximate production (m³/hr)				
Large machine	550	450	350	200
Medium machine	350	275	200	—
Small machine	100	75	50	—

Figure 8.3 A wheel dozer.

material because it casts aside the excavated material while maintaining a constant forward movement.

8.5 GRADER

A motor grader (Figure 8.4) is used for the profiling and spreading operations so that the final accurate elevation requirements for the subbase, earthworks, or whatever activity is involved can be achieved. The machine is also frequently and usefully used for maintaining haul roads by reducing the irregularities in the surface and removing mud. This grading operation very effectively reduces the rolling resistance of the machines using the haul road. The use of graders can therefore be extremely economical for overall earth moving production.

Graders can travel between 25 and 45 km/hr and work at speeds of 3 to 10 km/hr. The blade can be offset by a hydraulic sideshift and can be lowered or raised by ±400 mm and produce an accuracy to the final earthworks elevation of ±6 mm. The width of the blade varies from 2.4 to 3.1 m depending on the size of machine.

Figure 8.4 A motor grader.

A bulldozer blade can be attached to the front of the machine when the grader is required for spreading and dozing operations and a ripper-scarifier can also be attached to break up asphalt or rip up rocky material or frozen ground.

8.6 HAULING EQUIPMENT

There are several types of equipment that solely transport materials on a civil engineering site. The two most common types are rear dump trucks and dumpers. Other types of machinery used to transport material are belt conveyors, aerial ropeways, forklift trucks, and hover equipment.

8.7 DUMP TRUCKS

Generally the most suitable machine that can be used for long distance hauling on civil engineering works, open pit mines, and quarries is the rear dump truck. The capacity of a rear dump truck is usually measured by the rated load. A correlation between the rated load, the struck, and heaped capacities is shown in Table 8.3. The rated load of the trucks usually varies from about 17×10^3 to 85×10^3 kg, but very large trucks can handle up to a rated load of 350×10^3 kg.

The method for determining the hourly production of rear dump trucks is similar to the method of determining the hourly production of scrapers in which the overall cycle time is calculated. The loading time largely depends on the capacity and number of cycles of the loader or shovel to fully load the truck. Obviously, the hourly production of a hauling unit partly depends on the hourly production of the loading unit. It is therefore essential to achieve a balance between the loading unit and the hauling units. A good balance is usually achieved if the

TABLE 8.3 Rated Loads and Capacities for Dump Trucks

	(kN)	170	220	250	350	450	650	850
Approximate rated load	(Tonnes) ($\times 10^3$ kg)	17	22	25	35	45	65	85
Capacity (m³)	Struck	8.5	11	12	17	23	32	36
	Heaped	10.5	13.5	15	21	28	40	47

loading machine requires from about three to six cycles to fill the hauling machine.

The Caterpillar Tractor Co.[13] uses the following fixed times for dump truck productivity evaluations:

Total exchanging truck time	0.6 to 0.8 min
Maneuvering at the dump and dump time	1.0 min
Load time with wheel loader	0.50 min/cycle
Load time with hydraulic face shovel	0.35 min/cycle
Load time with cable operated power shovel	0.40 min/cycle

The actual times vary from those given above according to the job and conditions prevalent on the site. The load times should not include the first cycle time, as this time is subsumed into the exchange time; that is, if a wheel loader fills a truck in five cycles, the total load time is (5 − 1) × 0.5 = 2.0 min. The dumping time depends on the type of dumping operation. For instance, dumping in a stockpile is quicker than gradually spreading a truck load.

Haul and return times can be estimated by determining the gear in which the truck can travel, hence its maximum speed. The approximate maximum speeds in certain gears for several different types of rear dump trucks are given in Table 8.4.

An example of the equivalent gradient versus speed curves that can be found in the handbooks supplied by the machine manufacturers is

TABLE 8.4 Dump Trucks—Maximum Speed Versus Gears

Gear	Approximate maximum speed (km/hr)		
1	6	9.5	8
2	8.5	14.5	12
3	16	20	16
4	31	27	23
5	48	38	31
6	—	56	48
Reverse	8	8	4.5
Approximate rated load (kN):	170	250	350
($\times 10^3$ kg):	17	25	35

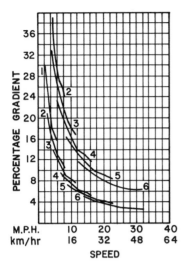

Figure 8.5   Effective gradient versus speed.

given in Figure 8.5. The lower lines represent the truck's performance when it is loaded and the upper lines represent the unloaded truck's performance.

Since the presentation of these graphs is not standard in the various equipment suppliers' handbooks, great care should be exercised when reading the graphs or tables, and values should be checked to ascertain which factors were considered by the manufacturer when the graphs were compiled.

If a dump truck is operating on a steep downhill gradient, the speed of the truck may have to be less than the speed deduced from effective grade versus speed graphs because the truck has to slow down to maintain a truck speed that the brakes can safely handle without exceeding the cooling capacity. The highest obtainable speed range within the braking capacity of the machine can be obtained from graphs of brake performance for gross weight versus effective grade versus speed given in the machine manufacturer's handbook.

8.8   DUMPERS

The small type of dumper truck is one of the most common and useful machines utilized on construction sites. It is used to transport materials that are required in small quantities. In the normal range of machines the bucket capacity varies from about 0.5 m³ to the large 3 m³ heaped type, or 0.7 Tonnes (0.7 × 10³ kg) to 5 Tonnes (5 × 10³ kg) in weight. The machines, which are either two or four wheel drive, are sometimes front tipping, but there are also turntable models which provide all round tipping. High discharge models are available which are suitable for discharging concrete into formwork above ground level.

8.9 BELT CONVEYORS

The troughed belt conveyor provides continuous delivery of material and can be an economic form of material transportation for short distances over routes that are likely to remain unaltered, which is the situation in most quarries. For long distances the initial cost of the plant is an important factor, but if large quantities of material are to be transported the belt conveyor becomes a feasible choice. Costs for conventional earth moving equipment rapidly increase with distance, whereas the increase in costs is not as sharp for the belt conveyor system. The operator cost element is low after the initial installation period.

Environmental advantages are also particularly worth considering today. Access roads to the site are quieter and cleaner and are not crowded with large earth moving trucks. A conveyor system can easily be bridged over roads, but the system must be enclosed to prevent spillage.

The system can be made up with certain standard length sections and with telescopic ends to permit greater flexibility. The length may be altered by the addition or removal of a standard section. Belt widths vary from about 0.5 to 1.0 m. Obviously the system works more economically when the loading and delivery areas are reasonably compact.

Material on the belt can travel at approximately 10 km/hr, and 2000 Tonnes (2000 × 10³ kg) of material can be moved per hour. The loading should be estimated at about 70% of maximum capacity to allow for irregular loading. The maximum gradients that the usual systems can handle is about 12° for gravelly rock and 20° for dry sand.

8.10 OTHER FORMS OF TRANSPORTATION EQUIPMENT

When material must be transported over difficult terrain or water, an aerial cableway can be used in preference to a haul road. The cost of installation may be economical when the loading and delivery points happen to be concentrated in small areas.

A forklift truck is suitable for fairly small volume loads that are packed together or stacked on pallets. It is particularly useful when the loads are to be transported and lifted into or from stacks. Loading or delivery points can be reached up to a maximum height of approximately 5.5 m.

Where ready mixed concrete is to be transported to site, concrete truck mixers are used. Mixer capacities are available in 4, 5, 6, 8, and 10 m³ volume or with agitator capacities of 5, 6.25, 7.5, 10, and 12.5 m³. Truck mounted mobile concrete pumps are capable of pumping up to 75

Plate 8.1   A standard 2 axle scraper. Courtesy Caterpillar Tractor Co., Peoria, Ill.

Plate 8.2   A twin engined scraper. Courtesy Terex, G. M., Hudson, Ohio.

Plate 8.3 A scraper pushed by tractor. Courtesy Terex, G. M., Hudson, Ohio.

Plate 8.4 A push-pull scraper. Courtesy Terex, G. M., Hudson, Ohio.

m vertically or 350 m horizontally at outputs of 32 m³/hr. An alternative to pumping, or skips and crane, for a large horizontal concrete pour such as a bridge deck is the self-propelled monorail system. The rail skips are side tipping, powered by gasoline engines, and as they run along the rail they may be stopped by a lever at the appropriate discharge position.

Machines using the hover principle exert extremely low pressure on

Plate 8.5 An elevating scraper. Courtesy Terex, G. M., Hudson, Ohio.

Plate 8.6 A D10 bulldozer. Courtesy Caterpillar Tractor Co. Peoria, Ill.

Plate 8.7 A bulldozer. Courtesy Terex, G. M., Hudson, Ohio.

Plate 8.8 A track type loader, dozing with multipurpose bucket. Courtesy International Harvester, London, England.

Plate 8.9 A loader in dozing operation. Courtesy Volvo Canada Ltd., Bromalea, Ont., Canada.

Plate 8.10 A grader. Courtesy Caterpillar Tractor Co., Peoria, Ill.

Plate 8.11 A dump truck (background). Courtesy Terex, G. M., Hudson, Ohio.

Plate 8.12 A coal hauler. Courtesy Terex, G. M., Hudson, Ohio.

Plate 8.13 A dumper. Courtesy Winget Ltd., Rochester, Kent, England.

Plate 8.14 Belt conveyors. Courtesy Eagle Iron Works, Des Moines, Iowa.

Plate 8.15   An aerial cableway. Courtesy Sauerman Bros. Inc., Bellwood, Ill.

the ground. This enables loads to be moved over ground with very low bearing pressures, such as bogs or snow and ice, buried pipelines, or weak structures that would require strengthening for high wheel loads. Apart from the standard hover trailer or platform, wraparound hover skirts are also available.

8.11   WORKED EXAMPLES

Example 8.1

Define:

(a) Rimpull
(b) Rolling resistance
(c) Traction
(d) Fixed time
(e) Grade resistance

in connection with earth moving scrapers.

Determine the weekly production (in in situ m³) of three rubber tired tractor scrapers, with a capacity of 20 m³ per scraper, when the soil is heaped.

Assume a 70 hr work week and no stoppages from major breakdowns or inclement weather. The level of the site varies between 116.09 to

125.71 m above sea level, so altitude does not affect the power of the scrapers.

Data

Soil properties	Moist clay, 17 kN/m³ (in situ), coefficient of bulking 0.40
Haul	720 m zero grade, 180 m 5% uphill
Return	1 km zero grade
Haul road	Firm, smooth, rolling resistance 3.25% (i.e., of gross weight)
Coefficient of traction	0.50
Efficiency	75% (i.e., 45 min/hr)
Fixed time	3:00 min
Weight of tractor and scraper	270 kN empty
Weight on drive wheels	190 kN loaded

Gear	1st	2nd	3rd	4th	5th
Speed (m/sec)	1.8	3.5	5.8	9.7	16.0
Available rimpull (kN)	97	50	30	18	11

Answer 8.1

(a) RIMPULL is the pulling force that the engine delivers to drive the tires at the ground contact point.

(b) ROLLING RESISTANCE is the resistance to movement of wheeled machines over level ground.

(c) TRACTION is the ability of the drive wheels or tracks to grip the ground.

(d) FIXED TIME is the part of the cycle time for machines taken in loading, dumping, turning, acceleration, deceleration, and so forth.

(e) GRADE RESISTANCE is the extra resistance to the movement of machines uphill, and therefore it increases the required tractive effort by approximately 1% for each 1% increase in the gradient.

Using equation 5.1 gives

$$V_l = (1 + C_b)V_i$$

$$V_i = \frac{20}{1 + 0.4}$$

$$= 14.29 \text{ m}^3$$

The total load carried per scraper per cycle = 14.29×17
$$= 242.8 \text{ kN}$$

Power requirements:
  Haul

$$\text{Rolling resistance} = (243 + 270) \times \frac{3.25}{100} = 16.67 \text{ kN}$$

$$\text{Grade resistance} = (243 + 270) \times \frac{5}{100} = 25.64 \text{ kN}$$

  Return

$$\text{Rolling resistance} = (270) \times \frac{3.25}{100} = 8.77 \text{ kN}$$

Therefore use a scraper in 2nd gear at 3.5 m/sec for 180 m and 4th gear at 9.7 m/sec for 720 m of the haul journey, and in 5th gear at 16.0 m/sec for 1 km of the return journey.

Power limitations:

$$\text{Altitude—not applicable}$$
$$\text{Traction:} \quad 190 \times 0.5 = 95 \text{ kN}$$

Note that the available rimpull is reduced from 97 to 95 kN in first gear. Generally a pusher tractor assists the scrapers to overcome slipping during the actual scraping operation.

  Cycle time

$$\text{fixed time} \qquad\qquad\qquad = 3.00 \text{ min}$$

$$720 \text{ m haul} = \frac{720}{(60 \times 9.7)} = 1.24 \text{ min}$$

$$180 \text{ m haul} = \frac{180}{(60 \times 3.5)} = 0.88 \text{ min}$$

$$1 \text{ km return} = \frac{1000}{(60 \times 16)} = 1.04 \text{ min}$$

$$\text{Therefore total cycle time} = 6.14 \text{ min}$$

$$\text{Corrected cycle time} = \frac{6.14}{0.75} = 8.18 \text{ min}$$

Production:

1 no. scraper, cycles/hr   = 60/8.18   = 7.34
1 no. scraper, production/hr = 14.29 × 7.34 = 104.8 m³ (in situ)
3 no. scrapers, production/hr = 3 × 104.8   = 314 m³ (in situ)
Therefore weekly production = 314 × 70   = 22,011 m³

Answer

22,000 m³ (rounded off) (in situ)

Example 8.2

This is an example in the determination of truck-shovel production.

Determine (a) the cycle time for the trucks
              (b) the number of trucks required
              (c) the production per month

in the following example.
     There are 769B Caterpillar trucks available. Assume that the rated load is 32 Tonnes (approximately 320 kN), the rolling resistance is 4%, the grade resistance is 2% (uphill) haul, and the efficiency is 50 min/hr. The haul distance is 1.5 km and an empty truck weighs 280 kN.

Truck data:

Gear	Maximum rimpull		Average speed for gear (km/hr)
	(kgf)	(kN approx.)	
1st direct	13,000	130	8
1st overdrive	10,000	100	11
2nd direct	6,000	60	16
2nd overdrive	4,000	40	25
3rd direct	2,500	25	40
3rd overdrive	1,600	16	60

The material to be hauled is sandy-gravel and a unit weight of 17 kN/m³ should be assumed. The fixed time for loading and unloading should be assumed to be 4.00 min.

Shovel data:

Bucket size $1\frac{1}{2}$ m³.

Production for sandy gravel = 250 m³, see Table 7.1, for optimum depth of 2.4 m and 90° swing. Assume that the correction factor for the shovel operating at 80% optimum depth and 45° swing is 1.22 and the carry factor for sandy gravel is 0.95. Use a job efficiency factor of 0.84.

An 8 hr day and 21 day month should be considered in the calculations.

Answer 8.2

(a) The heaped capacity of a truck with a 35 Tonnes (350 kN) rated load is given in Table 8.3 as 21 m³, and this value may be assumed to be close enough.

The rimpull required is

$$\begin{aligned}
\text{Empty} &&&= 280 \text{ kN} \\
\text{Full load} &= 21 \times 17 &&= 357 \text{ kN} \\
\text{but rated load} &&&= 320 \text{ kN}
\end{aligned}$$

Power requirements:

$$\text{Haul, rimpull} = (280 + 320)\frac{6}{100} = 36.0 \text{ kN}$$

Therefore the truck will travel in 2nd overdrive gear at an average speed of 25 km/hr.

$$\text{Return, rimpull} = 280 \times \frac{2}{100} = 5.6 \text{ kN}$$

Therefore the truck will travel in 3rd overdrive gear at an average speed of 60 km/hr.

Cycle time:

$$\text{Loading + unloading time} = 4.00 \text{ min}$$

$$\text{Haul time} \quad = \frac{60}{25} \times 1.5 = 3.60 \text{ min}$$

$$\text{Return time} \quad = \frac{60}{60} \times 1.5 = \underline{1.50 \text{ min}}$$
$$\underline{9.10 \text{ min}}$$

(b) Average shovel output $= 250 \times 1.22 \times 0.95 \times 0.84$

$$= 244 \text{ m}^3$$

$$\text{Required no. of truck trips/hr} = \frac{244}{21} \times \frac{357}{320}$$

$$= 12.96$$
$$= 13$$

Note that to obtain an average production for the shovel the job efficiency factor of 0.84 was included in the calculations. A 50 min hour is therefore assumed for the trucks.

$$\text{No. of cycles for each truck/hr} = 5.49 \text{ cycles/hr/truck}$$

$$\text{Therefore no. of trucks required} = \frac{13}{5.49} = 2.36$$

$$= 3 \text{ trucks}$$

One additional truck is often provided for every five trucks that are required.

(c)   Monthly production of shovel $= 244 \times 8 \times 21$
$$= 40{,}992 \text{ m}^3$$
$$= 41{,}000 \text{ m}^3$$

Example 8.3

Determine the monthly production if only two trucks are provided for the operation described in Example 8.2.

Answer 8.3

The production does not depend on the maximum output of the shovel, as the shovel is underutilized with only two trucks working.

If no breakdowns are assumed, the production of the two trucks

$$= 2 \times 5.49 \times 21 \times 8 \times 21 \times \frac{320}{357}$$

$$= 34{,}723 \text{ m}^3$$

$$= 35{,}000 \text{ m}^3/\text{month}$$

Note that it should not be assumed that the rated load can be exceeded when the haul journey is, on average, uphill.

Example 8.4

The graphs distance versus time loaded, and distance versus time unloaded are reproduced from the Caterpillar performance manual; refer to reference 13. Determine the variable times for the truck described in Example 8.2 using the graphs.

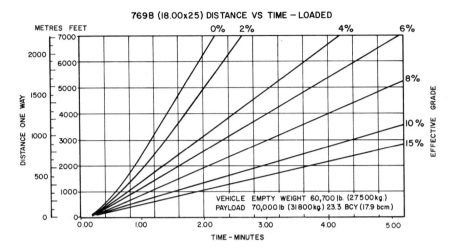

Example 8.4

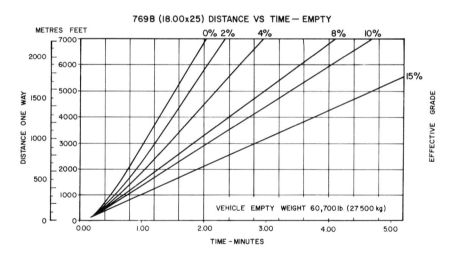

Example 8.4

Answer 8.4

Reading horizontally along the 1500 m distance until the 6% effective grade line intersects it for the (loaded) truck and the 2% effective grade line intersects it for the (unloaded) truck, the following times, in minutes, can be read by following the intersection point vertically downward:

Haul time = 3.60 min
Return time = 1.70 min

Example 8.5

The table given in Example 8.2 only gives the average speed for a given gear whereas the graph of rimpull versus speed more accurately

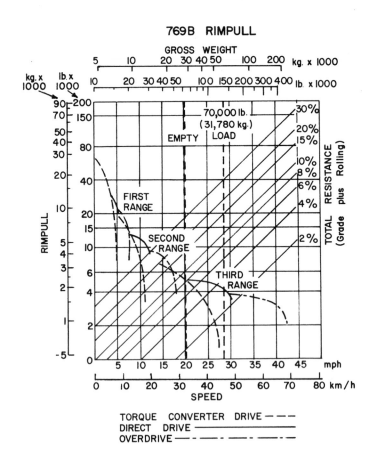

769B RIMPULL

TORQUE CONVERTER DRIVE ———
DIRECT DRIVE ——————
OVERDRIVE —— - —— - —— - —— - —

describes the ranges of rimpull and speed. Determine the haul speed of the truck described in Example 8.2, but use the graph of rimpull versus speed. Exactly the same conditions should be assumed.

Answer 8.5

Read from the gross weight of 60,000 kg (28,000 kg empty weight + 32,000 kg) down to the 6% (gradient plus rolling resistance) total resistance. From this weight-resistance intersection point, read horizontally to the curve with the highest obtainable speed range. This second intersection point gives the maximum available rimpull by reading horizontally to the vertical rimpull axis, and gives the speed of the truck by reading vertically to the horizontal speed axis.

In this particular example the speed is found to be 26 km/hr, in 2nd overdrive gear, compared with 25 km/hr in Example 8.2.

Example 8.6

To maintain a haul road properly, a motor grader must make one pass in 2nd gear (speed 6.0 km/hr) and then two passes in 3rd gear (9.2 km/hr). Determine the time to complete the job if the length of the haul road is 10.5 km.

Answer 8.6

Gear	Speed	Time taken	
2	6.0	$\dfrac{10.5}{6.0}$	$= 1.75$ hr
3	9.2	$\dfrac{2 \times 10.5}{9.2}$	$= 2.28$ hr
Total time			$= 4.03$ hr

Assume that the motor grader averages 50 productive min/hr, that is, a job efficiency factor of 0.83.

Therefore total actual job time $= \dfrac{4.03}{0.83} = 4.86$ hr

CHAPTER 9

COMPACTION

9.1 INTRODUCTION

Compaction is the process of packing soil particles more closely together by rolling or by other mechanical means. During this process the density of the soil is increased and the porosity is reduced.

An embankment must be sufficiently strong to support its own weight and superimposed loads. By increasing the density of the soil, the compaction process increases the shear strength, the impermeability, and the bearing capacity of the soil. Refer to Figure 9.1.

The future purpose of earthworks determines the degree of compaction required during its construction. Sometimes compaction equipment is not necessary. For an embankment where a solid pavement is to be constructed, or where fill is required under a slab foundation, settlement cannot be permitted and compaction equipment is therefore used to obtain a high degree of compaction. To obtain this high degree of compaction throughout the entire depth of the earthworks, the fill is usually constructed in layers ranging from 100 to 400 mm compacted thickness with a suitable type of compaction machine passing over each layer from 4 to 12 times. To achieve optimum conditions it may be necessary to adjust the moisture content of the fill during the earthworks construction.

Although the use of compaction equipment increases the initial costs, subsequent maintenance is reduced and the risk of slips is eliminated. The structural engineer specifies that fill, when it is required, is well compacted so that an appropriate bearing capacity may be assumed in design for the soil supporting the foundations of a permanent structure. The highway engineer also requires fill used for roads and runways to achieve sufficient strength to support the loads superimposed on the fill.

Over the last few decades compaction specifications have become

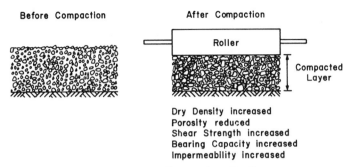

Before Compaction   After Compaction

Dry Density increased
Porosity reduced
Shear Strength increased
Bearing Capacity increased
Impermeability increased

Figure 9.1   Compaction of soil.

more precise. This has caused bottlenecks on the fill areas and therefore faster self-propelled compactors, of the smooth wheel or tamping
foot varieties, are now used to increase production. However, not all
compaction equipment has become larger and faster. If a roller is made
too heavy in an attempt to deepen the effective compacted depth, it
may tend to become bogged down on the initial pass. Heavy loads on
granular fill may crush the surface and cause stratification. Also, compaction may not be so effective as the roller speed increases.

During the initial pass of a smooth wheel roller on an uncompacted
layer of asphalt considerable penetration occurs. After successive
passes the pressure bulb depth decreases and high stresses occur at
the surface which sometimes causes the aggregate to crack. Pneumatic tired rollers were introduced to prevent the aggregate from
cracking and to simulate traffic conditions more closely.

9.2   FACTORS INFLUENCING THE METHOD OF COMPACTION

The three factors which influence the choice of the most effective compactor are as follows:

1.   The type of soil and distribution of particle sizes
2.   The moisture content
3.   The compactive effort

The compactive effort on the soil may be produced by a combination
of one or more of the following:

1. Pressure   The application of the weight of the machine
                           per unit area when stationary.

2. Impact   The energy that produces a shock load of short
                             duration and low frequency.

3. Vibration   The load produced by the vertical motion of the
                             compactor. Generally a separately driven vi-
                             brating unit produces the high frequency vi-
                             brations.

4. Kneading action   The porosity is reduced by the working of the
                             soil between the feet of the machine.

In considering the pressure exerted on a soil by the weight of a roller
and the variation of pressure against depth, it is convenient to consider
pressure bulbs produced from a loaded circular plate as shown in Figure 9.2a. If the same pressure q acts over a greater area, the depth of
the pressure bulb increases. An increase in the pressure also increases
the depth of the pressure bulb. Consideration of the pressure bulb
shows why in most specifications the maximum depth of a layer of fill is
specified. The categories in the specifications for the ranges of weights
of rollers also depend on the depth of the pressure bulb. The pressure
bulbs are shown for a roller in Figure 9.2b. The depth of a pressure
bulb increases according to the depth of penetration, which in turn
depends on the number of passes and the distribution and size of the
soil particles.

When the soil is loosely packed, compaction methods can improve
the bearing capacity to a maximum depth of about 400 mm. However,
if the soil is wet and the groundwater table is above foundation level,
the water table must be lowered by some means of drainage to improve
the allowable bearing capacity. Where effective drainage already
exists in sandy or gravelly soils, the use of compaction equipment,
which is effective to depths of up to about 300 mm for this type of soil,
can improve soil properties, but usually for clays and silts compaction

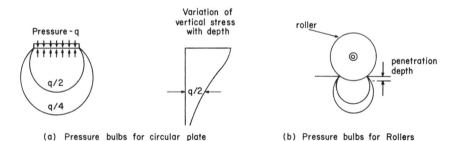

(a) Pressure bulbs for circular plate   (b) Pressure bulbs for Rollers

Figure 9.2 Pressure bulbs in soil.

is ineffective for the foundation of buildings, particularly in wet conditions. Naturally occurring soils or existing old fills, which require compaction but which have not been compacted in layers, may be compacted by deep compaction methods such as driving piles, the use of sand drains, or vibroreplacement methods provided that the site is large enough to warrant the expenditure. In the vibroreplacement method vibration equipment usually consists of an eccentrically mounted weight rotating inside a tube, which is usually lifted into position by a crane. The tube usually sinks down into granular soil under its own weight while the vibrator is operating, and as the vibrating unit is lowered a cylindrical void is formed in the soil, after which a clean sand-gravel mixture is added to fill the void.

9.3 MATERIALS

Most of the information in this section, including Table 9.1, has been obtained from the 'Specification for Road and Bridge Works,' HMSO 1976.[14] Suitable material used for embankments or other areas of fill is defined in Section 14.2. The definition of cohesive soils used in Table 9.1 includes clays and marls with up to 20% gravel or rock and with a moisture content not less than the value of the plastic limit minus 4. Well graded gravel-sand mixtures and dry cohesive soils include clays and marls containing more than 20% gravel or rock, or having a moisture content less than the value of the plastic limit minus 4, well graded sands and gravels with a uniformity coefficient exceeding 10, shales, and clinker ash. The uniformly graded material category listed in Table 9.1 represents sands and gravels with a uniformity coefficient of 10 or less, all silts, and pulverized fuel ashes. The uniformity coefficient is the ratio of the maximum size of the smallest 60% of the sample to the maximum size of the smallest 10% of the sample.

Materials for subbases and road bases include granular material such as crushed rock, crushed slag, crushed concrete, and naturally occurring well graded gravel-sand mixtures lying within certain grading limits. Cement-bound granular, soil-cement materials, wet-mix macadam, and other road base materials are also included in the group.

Fill material should be compacted as soon as possible after deposition, and according to the details given in Table 9.1. The engineer is entitled to arrange comparative field density tests to determine whether the compaction process has been carried out satisfactorily. The depth of the compacted layer, given in Table 9.1, refers to the

TABLE 9.1 Compaction Requirements—Maximum Depth/Minimum Number of Passes

| Compaction machine | | Maximum depth of compacted layer (mm)/minimum number of passes | | | Subbases and road bases |
| | | Earthworks | | | |
Type	Weight per meter width	Cohesive soils	Well graded, gravel-sand mixtures, dry cohesive soils	Uniformly graded material	Granular, cement bound granular, and soil-cement materials
Smooth wheel roller	up to 53 kN/m	125/6	125/8	125/8^T	110/16
	over 53 kN/m	150/4	150/8	UNS	150/16
Grid roller	up to 53 kN/m	150/10	UNS	150/10	UNS
	over 53 kN/m	150/4	150/12	UNS	UNS
Tamping roller	over 50 kN/m	225/4	150/12	250/4	UNS
Vibrating roller	up to 18 kN/m	125/8	150/8	200/10^T	150/16
	18 kN/m up to 42 kN/m	225/4	225/4	300/8^T	225/7
	over 42 kN/m	275/4	275/4	300/4^T	225/5

Pneumatic tired roller	Weight per wheel				
	up to 25 kN	175/4	125/12	UNS	110/12
	25 kN to 59 kN	300/4	125/10	UNS	110/12
	59 kN to 117 kN	400/4	150/8	UNS	150/16
	over 117 kN	450/4	175/6	UNS	150/12
Vibrating plate compactors	Weight per m²				
	up to 14 kN/m²	UNS	75/6	150/6	UNS
	14 kN/m² to 21 kN/m²	150/6	150/5	200/4	150/8
	over 21 kN/m²	200/6	200/5	250/4	
	Weight				
Vibrotamper	up to 0.75 kN	125/3	125/3	200/3	225/10
	over 0.75 kN	200/3	150/3	225/3	225/8
Power rammer	up to 5 kN	150/4	150/6	UNS	150/8
	over 5 kN	275/8	275/12	UNS	225/12

T—Rollers must be towed; self-propelled rollers are unsuitable. UNS—unsuitable.

finished depth of the material after it has been finally and satisfactorily compacted. The number of passes refers to the number of times a particular area of fill has been completely traversed by the compaction machine.

As Figure 9.3[15] shows, an increase in the number of passes that a soil receives increases the dry density or dry unit weight up to a certain moisture content. However, if too much rain falls on an uncompacted clay fill, it may be impossible to achieve the required degree of compaction until the fill has dried out. Plowing up the surface helps the evaporation process but, compared with compacted fill, uncompacted fill takes longer to dry out after rain or a snowfall. It is for this reason that it is important to use compaction plant as soon as possible after deposition of the fill. During hot spells it may be necessary to add water to achieve the required degree of compaction.

9.4 CHOICE OF COMPACTION EQUIPMENT

Table 9.1 serves as a useful general guide for the suitability of the various types of compaction equipment available for soils within the three categories of suitable materials, road base, and subbase materials. Figure 9.3 is a typical graph of dry unit weight against moisture

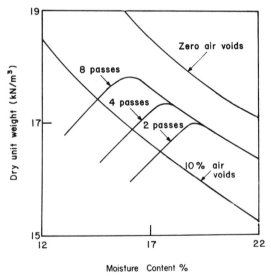

Figure 9.3 Dry unit weight (density) versus moisture content.

content for a particular soil and compaction machine. It can be seen that there is a limit to the usefulness of the number of passes and that it is not worthwhile increasing the number of passes if the soil possesses a certain limiting moisture content.

For many years compaction specifications have been based on a technique involving the relative density or degree of compaction. In the almost ideal condition of a soil mechanics laboratory a certain density can be achieved in standard tests, but on construction sites the same density is difficult to achieve for the same soil. Therefore to allow for this variation it was usual to relate the specified density, which could reasonably be expected to be achieved on site, to the maximum dry density achieved in a standard compaction test. Compaction specifications have more recently included the method of compaction, that is, the maximum depth of the compacted layer and minimum number of passes for one type or range of compaction machine. This type of specification, sometimes referred to as the method procedure, uses a table similar to that shown in Table 9.1 and usually allows the contractor alternative procedures, not given in the specifications, provided that site trials are conducted for the engineer before approval.

Although the use of a table similar to Table 9.1 is usually regarded by the engineer as a better method of choosing compaction equipment and deciding the number of passes required, there are often strong economic arguments put forward by the contractor that an end test result procedure is better than a method procedure. It can certainly be shown that for borderline cases an anomalous situation arises with the method procedure, where a slight increase in the weight of a vibrating roller, for instance, would decrease the actual end result because the required number of passes can be reduced from 16 to 8.

Alternatively, shear strength or maximum air void content test procedures with a controlled moisture content may be specified. Dual specifications may also be given which include an end test procedure in conjunction with the method procedure.

Smooth wheel and pneumatic rollers are primarily intended for road surfacing and road bases. The maximum depth of the compacted layers is restricted to shallow depths of about 100 to 150 mm. For large rock fills a large heavy vibrating roller is most suitable. It is possible to compact layers up to depths of about 1 meter.

Gravels or finely graded rock fill may be compacted efficiently by tamping or vibrating rollers. The tamping roller compacts a comparatively shallow layer up to about 250 mm at speeds up to 12 km/hr. The vibrating roller compacts a comparatively thick layer of 500 mm to 1000 mm at a speed up to about 4 km/hr. The vibrating roller is therefore more compatible with dump trucks, which tip material in thicker layers, than with scrapers. Vibrating rollers can also be used on other types of soils but are generally less economical than tamping rollers.

The grid roller was originally developed to provide cheap, unpaved, all weather roads by crushing the large lumps of rock and filling the voids with the broken finer material. It is particularly effective on granular soils, but the more plastic soils, such as some clays, tend to clog the grids and the roller becomes ineffective.

For sands and silts tamping rollers are most suitable, but they do not perform well on loose material such as dune sand. The tamping roller eliminates stratification and also prevents the formation of a wave of material in front of the roller. It is a versatile and economical machine and capable of compacting most common soils.

In clay and boulder clay compaction is achieved by kneading the particles together and thus reducing the air voids. High speed self-propelled tamping machines are usually most effective in good weather conditions, but in inclement weather water is held in the cavities produced by the feet, creating a soggy surface. Obviously the configuration of the tamping roller inherently produces excellent traction. Pneumatic tired rollers with ballast are occasionally used successfully and towed by a crawler bulldozer to increase traction. Towed vibrating rollers can also be used to produce a smooth surface that assists in the drying process after wet weather. Organic silts and clays are unsuitable for fill material. Compaction is impracticable with peat and other highly organic soils, and this type of soil should be removed for load bearing construction.

Hard chalk may be considered as hard rock in the selection of compaction equipment. Care should be exercised not to overcrush soft chalk, and therefore a pneumatic tired roller, where tire pressures can be adjusted for optimum pressures, is the most effective roller for soft chalk.

Compactors achieve their maximum effect in the first few passes and for each successive pass the rate of increase in density or unit weight of the fill is progressively less. The usual recommended number of passes for the maximum depth of compacted layer of a particular soil type is given for each type of compaction machine in Table 9.1.

Although the unit cost of compacting a cubic meter of soil may be small compared to the unit costs of loading, hauling, and spreading, substantial economies can be made by choosing the correct machine for a particular job. The number of passes, the effective width of the machine, the loose depth of each layer of fill, and the machine's operating speed all contribute to the rate of production, and hence the correct choice of the compaction machine. The choice of compaction machine also depends on the availability of machines, the size and duration of the job, the type of soil, the layout of the job, its compatibility with the proposed earth moving equipment, the program of work, the capital cost, and the estimated effective utilization of a machine during its life.

9.5 DRY DENSITY/MOISTURE CONTENT RELATIONSHIP

The bulk density of a soil is the mass of material, including any water and voids, per unit volume. The dry density or dry unit weight of a soil is the mass contained in a unit volume of undried material after drying at 105°C. The moisture or water content is the mass of water that can be removed from a soil by heating at 105°C, expressed as a percentage of the dry mass.

If water is added to a dry soil sample, the unit weight to which the sample can be compacted increases to a point termed the optimum moisture content and then decreases as more water is added. This effect is shown in Figure 9.4 on all the curves.

The optimum moisture content value may occupy a zone over a range of moisture content, as shown in curve A, or the curve may abruptly change direction as shown in curve B. This difference depends on the type of soil. To establish the optimum moisture content for a soil at the time of compaction, a series of laboratory standard compaction tests are conducted with progressively increasing quantities of water, to

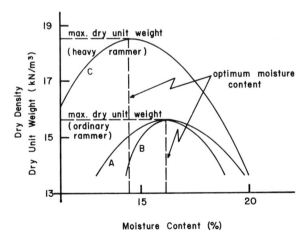

Figure 9.4 Dry unit weight versus moisture content.

produce a graph similar to the one shown in Figure 9.4. The peak of the curve on the graph gives the dry unit weight/optimum moisture content relationship.

9.6   SOIL COMPACTION TESTS

The dry density/optimum moisture content relationship does not have a unique characteristic value for a particular soil because the position of the peak on the graphs can vary depending on the applied compactive effort. If a heavy rammer is used, refer to curve C in Figure 9.4, a different optimum moisture content may be obtained for the same soil as that which produces curve A.

The first well known compaction test used to determine the dry density/optimum moisture relationship was the Proctor test. This test was intended originally for the smooth wheel medium weight rollers available at the time. As the rollers became heavier, it was realized that to simulate site conditions more accurately a test with a higher compactive effort would be necessary to correlate site and laboratory data and so another test was introduced. An additional test has also been introduced for the granular type of soil, which is compacted by a vibrating roller on site. This test, which is described later, specifies the use of a vibrating hammer to simulate the effects of vibration.

Various national standards[16-18] describe a soil compaction test in which the dry density/moisture content relationship is determined. A metal rammer weighing 2.5 kg with a 50 mm diameter circular face is used.

A sample of air dried soil passing a 20 mm sieve is placed in a standard mold with an internal diameter of 105 mm. One third of the amount of soil required to fill the mold is given 25 blows (27 blows in some standards) from the rammer, which should be dropped from a height of 300 mm. The blows should be uniformly distributed over the surface of each layer. The process is repeated for the other two layers required to fill the mold. Suitable increments of water are added successively to alter the moisture content. At least five determinations should be made. Generally for sandy and gravelly soils a moisture content beginning at 5% with increments of 1 to 2% are suitable, and for cohesive soils it is recommended that the moisture content should begin at about 8 to 10% below the plastic limit of the soil, with increments of 2 to 4%.

An alternative compaction test, using a 4.5 kg rammer dropped through a height of 450 mm, is used to more accurately simulate the heavier machines used on site. A larger mold is used in the test and tamping is done in five layers of approximately equal volume. Small amounts of stone retained on a 20 mm test sieve may be removed from the sample, generally without any appreciable alteration to test results, but calculations should be made to assess the effect of the removal of large stones on the dry density/moisture content relationship.

For the determination of the dry density/moisture content relationship of granular soils passing a 37.5 mm sieve an electric vibrating hammer is used with a total downward force of between 300 and 400N to more accurately simulate the effects of a vibrating roller on site.

The determination of the in situ dry density or dry unit weight is usually carried out by a sand replacement method on construction sites. A metal tray with a hole in it approximately 100 to 200 mm in diameter is placed on a flat part of the fill to be tested. This hole is used as a template to dig a hole to the depth of the layer to be tested, up to a maximum depth of 250 mm. A clean dry uniformly graded natural sand is allowed to fall from a pouring cylinder which is weighed before and after use, and the volume of sand that fills the hole and a cone left at the top may be deduced. The weight of the sand required to fill the excavated hole is therefore equal to the weight of the cylinder and the sand before filling the hole less the weight of the cylinder and sand after filling the hole with an adjustment made for the mean weight of sand in the cone. By knowing the bulk density of the sand the volume of the hole may be calculated, and by weighing the mass of the soil excavated before and after drying, the dry density or dry unit weight and the moisture content can be deduced. Increasing use is now made of nuclear densometers, using backscatter techniques, to enable the degree of compaction to be measured rapidly. Also, for vibrating compactors the electronic compaction meter is a new measuring device that receives signals from an accelerometer which continuously records the vibrations of the drum while the roller is working. A special miniprocessor analyzes the signals, and the meter shows relative values that correspond to the dynamic modulus of elasticity of the soil, which enables the corresponding bearing capacity to be evaluated. The meter can be preset to a value corresponding to the specified degree of compaction obtained from one of the conventional compaction tests.

9.7 COMPACTION OF AREAS OF FILL

After suitable material has been placed in embankments or other areas of fill, it should be compacted as soon as practicable in layers with depths in accordance with the specification. Embankments should be built up evenly in layers just greater than the final full width of the

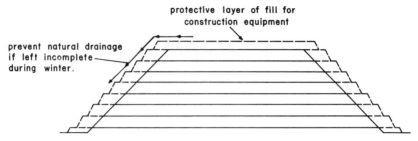

Figure 9.5 Construction width of embankments.

embankment, as shown in Figure 9.5, with sufficient slope and surface regularity to enable rain water to drain away rapidly.

If the embankment is incomplete during the winter months, natural drainage down the side slopes should be prevented. When originally suitable material deteriorates so that it does not comply to the specifications, it should be removed or made good. This procedure is very expensive, so every effort should be made to prevent the material from deteriorating.

Fill material is usually spread and leveled by a crawler tractor, often with the help of a grader, and the top surface side slopes are thoroughly blinded, where necessary, with finely graded material when rock fill is used. This material may be top soil. Isolated boulders may be included in the fill, subject to the requirements of the specifications.

If construction equipment is required to be used on the embankment, a protective layer of fill should be laid from about 100 to 300 mm in depth. Special methods using smaller compaction equipment, often pedestrian controlled, are required around the vicinity of pipes, subways, or culverts which traverse the area of fill.

The compaction specification for the soil types used in embankments is often a method and machine specification similar to that shown in Table 9.1. In addition the engineer may often check the method and machine specification by comparative density tests. Other types of specifications suggest a particular method and machine. Sometimes comparative density tests only are specified, leaving the contractor the choice of equipment.

9.8 COMPACTION MACHINES

The following sections describe the most common types of compaction machines[19] that are available for use on a construction site. The method of compaction for a particular operation, the type of fill, and all the other criteria mentioned in Section 9.4 determine the type of machine to be used.

The smooth wheel rollers rely on the static weight of the roller to exert pressure on the fill to achieve compaction. When kneading action is required on a particular soil, grid, tamping, or pneumatic tired rollers are suitable. Vibratory rollers produce a load by the vertical motion of a separately driven vibratory unit. Vibrating plate compactors, usually manually controlled, produce the same action on the soil and are useful for small or confined areas. Another compactor that is suitable in small or confined areas is the mechanical punner, which incorporates a rammer and produces a shock load of short duration and low frequency on the soil.

Self-propelled compactors, with relatively high working speeds up to about 10 km/hr, are becoming widely used on sites. These machines consist of part of a tractor unit including the engine and rubber tired wheels, and at the other side of an articulated steering joint is a vibrating roller, usually of the smooth wheel or tamping type.

9.9 SMOOTH WHEEL ROLLER

Smooth wheel rollers (Figure 9.6) are of the conventional smooth steel tandem or three wheel drum types, which are slow and usually used for road surfacing in shallow lifts of 50 to 150 mm. A roller is usually selected in terms of its weight per meter width of roll because of the great variation in weights. The compactive effort is a function of the weight of the machine. In Table 9.1 a roller is assessed on the axle that has the highest weight per unit meter when it has more than one axle.

As each pass of the roller is made, the contact pressure increases and the pressure bulb decreases. This effect can cause stratification of the

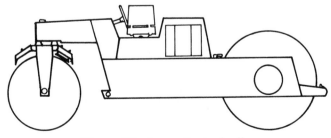

Figure 9.6 A smooth wheel roller.

layers, if a crust is formed on the surface of each layer, and can also crack the aggregate. The weight distribution can be altered by adding sand, water ballast, or ballast weights. The steel roll initially forms a wave in front of the roll, but the kneading action is lost as there is no weight on the wave. The roll also tends to bridge over soft spots. Some types have front and rear roll steering to facilitate compaction on horizontal curves.

9.10 PNEUMATIC TIRE ROLLER

Pneumatic tire rollers (Figure 9.7) are suitable for compacting cohesive and well graded granular soils and give a kneading or manipulating action. They are very effective on the surface courses of highways and are frequently used on road base and subbase courses. Towed pneumatic tire rollers with total weights from 300 to 1000 kN provide an excellent kneading action on embankments, and the compactive effort can be effective up to about 500 mm. The high loads can cause excessive penetration of the surface, which increases the rolling resistance and causes slow speeds and the requirement of powerful tractor units.

Self-propelled pneumatic tire rollers usually consist of two axles with multiple wheels, the back wheels being out of track with the front wheels so that the whole width is covered with some overlap. To induce further kneading action and to prevent bridging over soft spots or depressions, the wheels can have a tilt axle oscillation action or a vertical oscillation action, but this is an extra feature and is not so necessary on surfaces where a hard underlayer exists. Compaction can be improved by front and rear axle steering or centerpoint steering.

The tires must have a wide flat tread to reduce rolling resistance and

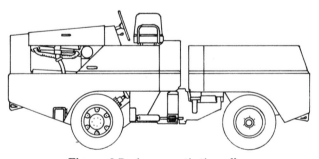

Figure 9.7 A pneumatic tire roller.

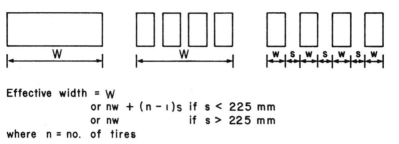

Effective width = W
          or nw + (n − 1)s if s < 225 mm
          or nw   if s > 225 mm
where n = no. of tires

Figure 9.8 Effective width of pneumatic tire roller.

improve compaction. Different types of soil mixes and materials require different tire pressures, and therefore a centralized rapid adjustment air control system, ensuring equal pressure in all tires, can be provided.

The weight per wheel is the total weight of the roller divided by the number of wheels. In Table 9.1 the effective width is taken as the sum of the individual wheel track widths and includes the sum of the spacing between the wheels provided that each individual space does not exceed 225 mm. If the spaces between the wheels exceed 225 mm, the effective width should be taken as the sum of the individual wheel track widths only for single axle rollers, as shown in Figure 9.8.

9.11   GRID ROLLER

The grid roller (Figure 9.9) is particularly effective for compaction and breaking down an unsurfaced all weather road where the fill initially contains larger lumps of rock and is generally drawn by a fast rubber tired tractor. The larger pieces of rock are broken down and fill the voids to produce an economical high speed unpaved road. The grid roller can compact a wide range of well graded granular soils at high

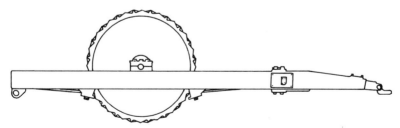

Figure 9.9 A grid roller.

speeds, and the structure of the grid creates a satisfactory kneading action on cohesive soils provided that the soil is not in a state that will clog the grids. The maximum depth of the compacted layer should be about 150 mm.

9.12 SHEEPSFOOT AND TAMPING ROLLERS

The towed sheepsfoot roller has been in use for many years. It has feet that project from the roll, as shown in Figure 9.10a.

Because of the fairly intense pressures at the tip of the foot, most of the compaction is produced by kneading or penetration. The sheepsfoot roller is useful in clays when stratification must be prevented, but its use has declined over the last decade.

A more recent development of a roller with projecting feet is the tamping roller (Figure 9.11). The foot is carefully designed to prevent the soil being thrown up when used at speed. The area of the foot increases with penetration so that the compactive effort is automatically adjusted. The size, shape, and spacing of the feet are important and various combinations are available. The tamping foot roller may be towed, but, as it becomes even more effective at higher speeds, a self-propelled tamping roller was developed. Tests have shown that an extremely high output may be obtained. Stratification is eliminated because of the kneading action, which takes place provided that the depths do not exceed 300 mm. About 200 m³/hr can be compacted in cohesive soils at a depth of 250 mm with about six passes. For uniformly graded soils about 240 m³/hr output is achieved with four passes in a depth of 250 mm. This machine can therefore prove to be extremely economical even though hourly costs for high speed compactors are considerably higher than the slower machines. Also, with the advent of larger and faster earth moving equipment the compaction operation has often been a bottleneck in a construction program, thus making self-propelled high speed compactors preferable.

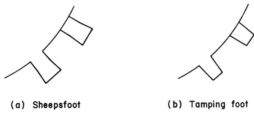

(a) Sheepsfoot   (b) Tamping foot
Figure 9.10 Sheepsfoot and tamping foot roller.

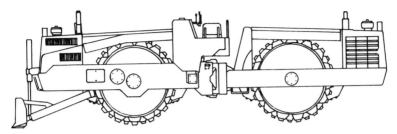

Figure 9.11   A tamping roller.

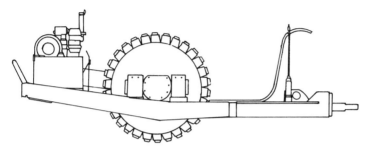

Figure 9.12   A vibratory tamping compactor.

A tamping roller is defined as a roller with projecting feet where the sum of the areas of the feet exceed 15% of the area of the cylinder described by the ends of the feet. The values given in Table 9.1 assume that the machines have two rolls in tandem; therefore if only one roll covers the surface on one pass, the number of passes should be multiplied by 2 in the table.

A vibratory unit may be attached to a towed tamping roller, as shown in Figure 9.12, but is only effective up to speeds of about 5 km/hr.

9.13   VIBRATING ROLLER

The vibrating roller (Figure 9.13) is usually a smooth wheeled machine with an engine driven vibratory unit attachment. They are produced in various sizes. Large vibrating rollers, with weights up to about 150 kN, are used on rock and gravel fills with a maximum speed of 5 km/hr. On other types of soils they are considered less economical than tamping rollers. Medium sized machines are used for granular materials and are useful in confined areas such as subways, culverts, abutment, and

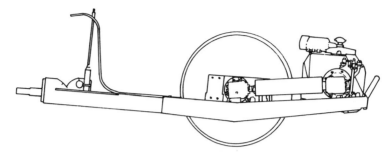

Figure 9.13 A vibrating roller.

wing wall areas. The smaller hand propelled machines are mentioned in Section 9.14.

Vibrating rollers temporarily reduce the internal friction of a granular soil, and therefore the soil particles are rearranged in a denser configuration. The effectiveness of a vibrating roller depends more on the energy transmitted into the soil than on the total weight of the machine.

The main advantage of vibrating equipment is the greater effective depth of compaction in granular materials. Some of the larger machines can operate at a high vibration frequency and small amplitude, which is suitable for near surface compaction, or at a low vibration frequency and a large amplitude, which is suitable for deeper compaction. Care should be taken in manufacture to ensure that the suspension of the vibrating roller isolates the vibrating forces, as transmission of these forces to the frame and engine leads to considerable inefficiency.

Where the vibration is applied by two rollers in tandem the minimum number of passes may be halved in Table 9.1. If one roller differs in weight per meter width, the smallest value should be taken if the two rollers are considered, or the highest value taken if one roller is considered.

9.14 SMALLER COMPACTION MACHINES

There are several small compactors that are hand operated, pedestrian controlled, and suitable in confined, stepped, or small areas. Small versions of the larger rollers such as the smooth wheel and vibrating roller are available for use on small or special areas that require the

Plate 9.1 A steel wheel compactor. Courtesy Hyster Co., Kewanee, Ill.

Plate 9.2 A smooth wheel vibrating roller. Courtesy Bomag (GB) Ltd., Eynsford, Kent, England.

Plate 9.3 A pneumatic tire compactor. Courtesy Hyster Co., Kewanee, Ill.

Plate 9.4 A tamping embankment compactor. Courtesy Hyster Co., Kewanee, Ill.

Plate 9.5 A towed tamping vibratory compactor. Courtesy Hyster Co., Kewanee, Ill.

Plate 9.6 A double drum vibrating compactor. Courtesy Hyster Co., Kewanee, Ill.

Plate 9.7 A towed vibrating compactor. Courtesy Hyster Co., Kewanee, Ill.

Plate 9.8 A small vibratory roller. Courtesy Bomag (GB) Ltd., Eynsford, Kent, England.

Plate 9.9   A small smooth wheel roller. Courtesy Winget Ltd., Rochester, Kent, England.

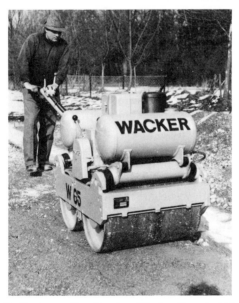

Plate 9.10   A small vibratory roller. Courtesy Wacker Canada Ltd., Mississauga, Ont., Canada.

Plate 9.11 A vibro plate. Courtesy Wacker Canada Ltd., Mississauga, Ont., Canada.

use of small machines. There are also small, special purpose compacting machines.

The power rammer contains a gasoline driven piston which causes the rammer to jump when fired. Compaction occurs on the firing stroke and also when the machine lands after the jump. A well balanced machine should enable the experienced user either to retain the machine in one spot when in use or to gradually move forward in any direction. Some machines are designed to jump forward automatically after each explosion. In wet cohesive soils energy is lost by suction, and the machines can become bogged down in clays of a highly plastic state, rendering them unsuitable in this situation. The equivalent of one pass is considered to have occurred when the shoe has dropped once on a given area.

Vibrating plate compactors contain a base plate to which a source of vibration is attached consisting of one or two eccentrically weighted shafts. They are normally operated at a speed of less than 1 km/hr, but if higher speeds are possible, the number of passes in Table 9.1 should

Plate 9.12   A vibro plate. Courtesy Wacker Canada Ltd., Mississauga, Ont., Canada.

be increased proportionally. The machines should be fairly lightweight with a low center of gravity to enable the machine to be easily maneuverable.

Vibro tampers contain a base plate in which oscillations are set up from an engine driven reciprocating mechanism.

9.15 PRODUCTION OF COMPACTORS

When a soil is compacted, it occupies a smaller volume because of the reduction of the air voids. The specification usually requires either that

Plate 9.13   Wacker rammers. Courtesy Wacker Canada Ltd., Mississauga, Ont., Canada.

a soil be compacted to a certain value or specifies the method of compaction or even requires a combination of both methods.

The compacted volume production P in m³/hr may be determined from the following equation:

$$P = \frac{W \times V \times D}{N} \text{ m}^3/\text{hr}$$

where W = effective compacted width per pass (m)
   V = velocity of the machine (km/hr)
   D = depth of compacted lift (mm)
   N = number of passes

Plate 9.14   Wacker rammers. Courtesy Wacker Canada Ltd., Mississauga, Ont., Canada.

The speed and depth of lift are of critical importance for a given machine. Some compactors, however, are only effective up to speeds of about 4 km/hr, but the special high speed compactors are effective up to speeds in excess of 16 km/hr. There is an optimum depth for the depth of lift (refer to Table 9.1), and production cannot therefore be increased by making the lift deeper for a particular machine. The approximate production of compaction machines can be determined by finding the specified compacted depth and using the effective speed of the machine, which can be obtained from the manufacturer's specification.

Plate 9.15 Wacker rammers. Courtesy Wacker Canada Ltd., Mississauga, Ont., Canada.

9.16 WORKED EXAMPLES

Example 9.1

Determine the weekly production of three rollers when the number of passes required in the specification is six, the average speed may be taken as 10 km/hr, and the compacted lift thickness is 150 mm. The overall width of the roller is 1.5 m.

Answer 9.1

Assume a compacted width of 1.4 m, a 35 hr week, and an hourly efficiency of 0.83. Then

$$P = \frac{W \times V \times D}{N}$$

$$\text{Hourly output/machine} = \frac{1.4 \times 10 \times 150 \times 0.83}{6}$$

$$= 290.5 \text{ m}^3/\text{hr}$$

Weekly output/three machines = $290.5 \times 35 \times 3$
$$= 30,502.5 \text{ m}^3/\text{wk}$$
$$= 30,000 \text{ m}^3 \text{ compacted/wk}$$
(rounded off)

Example 9.2

Estimate the hourly production of two towed pneumatic tire rollers with a single six wheel axle. Each wheel is 250 mm wide and 75 mm apart. Use Table 9.1 to obtain the requirements for compaction and assume an average rolling speed of 8 km/hr.

Answer 9.2

Use the equation in Section 9.10, Figure 9.8. Effective width is

$$W = nw + (n - 1)s$$

where n = number of tires
 s = distance apart
 w = width of each tire
 $= 6 \times 250 + 5 \times 75$
 $= 1.875$ m

Using Table 9.1, assume that the depth of the compacted layer is 110 mm and that the number of passes is 12, with an hourly efficiency of 0.83.
 Production of two rollers:

$$P = \frac{2 \times W \times V \times D}{N}$$

$$= \frac{2 \times 1.875 \times 8 \times 110 \times 0.83}{12}$$

$$= 228 \text{ m}^3 \text{ compacted/hr}$$

CHAPTER 10

EQUIPMENT ACQUISITION, FINANCE AND COSTS

10.1 INTRODUCTION

The method of determining the optimum time to replace equipment is extremely complex. About 25% of a construction company's capital is frequently invested in machines. In the past management by experience and instinct generally produced fairly successful results. Today a knowledge of a system that evaluates depreciation costs, maintenance costs, and other operating and ownership costs is important. Calculations can never be exact in the area of plant management, but the accurate recording of costs on equipment is essential so that a decision to replace a machine can be made at some time in the future, the decision being mainly based on a factual record of the machine's cost. There are complexities and anomalies in any system of plant management; for instance, a machine with an expected life span of 6 years may perform very satisfactorily for the first 3 years but may be unsatisfactory for the next 3 years, where the only difference, apart from age, is the ground condition encountered on the different sites.

The appropriate machines should be selected for specific site conditions to maintain a good performance, but unless the contractor has an almost unlimited choice of various machines it is not always possible to supply the most suitable machine for a particular site condition. For any given site condition, however, it is possible to manage machines so that their operating, maintenance, and repair costs are kept to a minimum with the maximum possible productivity and return on investment.

If a machine is managed properly, it should be inspected regularly to determine its actual condition. The time taken and the costs for maintenance and repairs should be recorded accurately so that some time in the future a medium term maintenance program and a long term replacement program can be developed.

Most contractors own about 75% of their total equipment requirements. The remaining equipment is hired, and can be effectively used during peak period requirements. Machines usually need to be fully utilized through their life for ownership to become a viable proposition. The fleet of machines owned by a company often influences the type of contract for which it is interested in bidding.

The civil engineer's involvement with plant acquisition for a contract begins during the tender stage. The estimator and planning engineer select the appropriate machines to perform particular jobs and, immediately before the tender is submitted, the directors may alter some of the initial machine selection, after consultations with the plant department, depending on the availability of certain machines. The project manager then makes the final machine selection at the beginning of the project in consultation with the site plant manager. The decision process is then based on many years of experience by several employees and the actual site conditions prevailing on site. The selection and acquisition of plant is a very important aspect of civil engineering, and a knowledge and interest in plant should be generated during a young civil engineer's academic and early professional training.

Construction equipment is available which offers versatility and greater productivity by using various features and attachments. Additional attachments should therefore be carefully considered so that the most suitable machine can be acquired for the long term. The availability of parts, service facilities, and backup support from dealers is also important.

10.2 PLANT EVALUATION AND SELECTION

Mechanization on construction sites has developed rapidly over the last few decades. The increasing cost of labor, technical advances in machines, and the expansion in plant hire have all contributed to the changeover to mechanization. Because of the number of similar machines that are often available, selection of the most suitable machine is difficult. The exact requirements should be listed, and technical and economical assessments should be made based on the manufacturers' data, company records, and the personal experience of plant personnel including the operators. The price, productivity, working dimensions, and special features and attachments are factors that are carefully scrutinized. Difficulties can arise when an attempt is made to compare the technical details of machines. The service facilities, maintenance requirements, and availability of machines and parts cannot be quantitatively measured.

A step by step systematic decision procedure should be developed in the evaluation and selection of equipment. First, the essential objectives should be isolated and established. Some weight should then be given to each objective so that it can be classified and listed in order of its relative importance. At this stage various types of machines should be compared with each other, and it may be necessary to compare the different sizes of machines from one manufacturer as well as machines from various other manufacturers. Any machine that does not conform to the obligatory objectives can be eliminated from the evaluation.

An extremely important factor of any machine is the initial cost. Great care should be exercised in comparing the initial costs of machines, and a distinction should be made between extra attachments that are included in the initial price and those attachments that are paid for separately.

Often when sensible systematic comparisons are made, one machine becomes the obvious choice. Other times the choice is not obvious and a decision is difficult. The performance and characteristics of the selected machine should be compared again with the second most favorable machine. Reevaluations should be made, particularly into the margins of error of the characteristics of the selected machine, and also into the disadvantages of the machine. When a final decision has been made to buy or hire a particular machine, the details of the second most favorable machine should be kept in case of unforeseen circumstances. If short term objectives differ appreciably from long term objectives for one machine, a decision to hire instead of buy will probably be made.

10.3 EQUIPMENT FINANCE

Expenses due to repairs and maintenance are greater than replacing a machine on a construction site after the machine has reached a certain optimum age. Therefore a time comes when a new machine must be purchased, although few companies today finance their plant fleet entirely through the cash flow of their own companies.

There are several methods of obtaining money over a medium term to finance new construction equipment. The financial lease, where the capital cost and the interest are recovered in a certain period, is gaining popularity. Major finance companies and banks finance the purchase of equipment, and the contractor requiring the equipment repays the money in installments over a certain specified period. Tax advantages usually go to the leasing company and not to the contractor, the lessee. When the repayment has been completed, the lessee may buy the equipment at a nominal cost or pay a nominal rent to the finance company, the lessor. The lessor therefore recovers the cost of the machine and interest over the term of the lease. An operational lease is a type of lease over a period less than the normal life of the machine and is accompanied by some service such as regular servicing or routine maintenance.

A loan may be arranged between a contractor and a bank to purchase equipment in much the same way as obtaining finance for a whole project. Interest rates are usually higher than the leasing method and fluctuate according to the bank or minimum lending rate. If new construction equipment is imported, favorable rates may be obtained from banks in the country of origin through some export incentive scheme.

Capital may also be obtained by the sale of ordinary shares, although steady growth and an expert's opinion are required. A much simpler approach is the use of an installment plan or hire purchase. This method of financing is often fairly good if the return on investment is in the short to medium term, but it becomes expensive in the long term. If interest rates are low and very short term financing is required, overdrafts from banks may be possible, but this method is limited. Higher than average interest rates are usually charged on overdrafts.

10.4 HOURLY COSTS

The contractor may obtain construction equipment for use on a particular job by either buying or hiring the equipment. To determine the cost per unit volume of moving the earth it is necessary to know the production and hourly cost of a machine. If the machine is hired, the hourly cost can be found from the schedule of rates of hire provided by the plant hire companies. The rates may be either for the machine only or an inclusive rate may be given where the rate includes operator's wages and overheads, although even these rates do not usually include fuel, insurances, overtime, or incentive payments. The estimator can, however, easily insert a cost per unit volume by calculating the production rate of the machine and using the hire rate with an allowance for company profit and so forth.

When large quantities are involved or when the continuity of machine operation is likely, it is probably more profitable for the main contractor to buy the machines, provided that there is enough capital available for the initial outlay. If the contractor owns the equipment, it is necessary to calculate an hourly ownership cost and an hourly operating cost. This cost can be fairly difficult to calculate accurately. Costs such as fuel may alter fairly rapidly with time, and costs can also vary considerably on a regional basis within one country. The hourly cost, however, can be of vital importance on the successful outcome of a bid. A difference of 10 cents per cubic meter can obviously make an appreciable difference in a competitive tender if 1 million cubic meters of soil are to be moved. An accurate hourly rate is therefore essential. A low estimate of the hourly rate would reduce or eliminate profits and a high estimate would lose the job. This particular rate is usually one of the most significant factors in the entire tender price of a contract containing large earth moving operations. Accurate data obtained from previous jobs enable the estimator to determine a more accurate hourly rate.

10.5 OPERATING COSTS

Operating costs include the cost for fuel, lubricants, filters, grease, and, when applicable, tire replacement and repairs, ripper tips and cutting edges, and so forth. Further costs incurred in operating a machine are repairs and labor and the operator's wage.

In deducing the hourly operating costs for fuel the contractor should refer to past company records and the fuel consumption data published by the machine manufacturer for the particular machine under consideration. The recommended servicing periods shown in the operator's manual indicate the average hourly cost of lubricants, filters, and grease in addition to any previous information logged in the company's records. The price of fuel can vary considerably from region to region.

Hourly costs vary according to the conditions prevailing on site. This is particularly true for tire costs. The haul road surface, maintenance, average speed and grades encountered, and wheel position all affect the life of a tire. The life expectancy of a tire can vary from 100 to 40% of the maximum tire life depending on whether conditions are favorable or otherwise.

Repairs and labor costs should be based on previous experience; however, a fairly common method of assessing repair costs, including new parts and labor, is to take a percentage of the 10,000 hr depreciation costs, even if the depreciation hourly cost is calculated over some other time period.

The operator's cost per hour should include the operator's wage rate, which varies regionally and can be obtained from various sources that publish the regional wage rates, and all overheads.

10.6 OWNERSHIP COSTS

Ownership costs include the cost of depreciation, interest, taxes, insurance, and storage. These are generally considered as fixed costs which the owner must pay and are not directly related to the annual utilization of the machine.

The most important unit cost for the ownership of a machine used on construction sites is depreciation. Usually a straight line graph is assumed from the new machine price to zero over the life expectancy of the machine. The scrap value of the machine at the end of its life is usually neglected. The declining balance method is sometimes used, which slightly more realistically assumes that depreciation is higher for the first few years of the machine's life. However, depreciation together with repair costs are expressed as an hourly rate throughout the life of the machine, and the straight line method is therefore usually assumed to achieve an equitable result. Typical values assumed in depreciation costs are 40 hr/week, 50 weeks per year and a 5 year life or greater depending on the type of machine. Depreciation is discussed in more detail in the following section.

All extra attachments, delivery charges, and so on are subsumed into the initial machine cost. Because tires are not expected to last as long as the machine, the tire costs are deducted from the initial machine costs and are regarded as an operating cost. The undercarriage of track type vehicles is, however, considered as part of the initial machine cost.

The interest on the overall cost of the machine, including delivery, must be considered even if the money required was not borrowed from an external source. An anticipated average interest rate should be used to calculate the interest on the total investment cost. A common method, which simplifies this estimate, is to assume a percentage of the actual delivery cost as an allowance for interest, insurance, property taxes, and storage. Insurance costs should cover all the fully comprehensive and liability policy premiums relating to the particular machine considered.

Example 10.1

The following worked example of machine costs is intended only as a guide. There can be considerable variation of costs, not just from one country to another but from one region to another within a country. Costs can also show a very considerable annual variation, particularly during periods of high inflation.

The estimated hourly ownership and operating cost of a rubber tired scraper may take the form of the following example:

Rubber Tired Scraper—Hourly Cost Estimate

$$\begin{aligned}
\text{Purchase price} &= \$200{,}000 \\
\text{Extras: canopy and so forth} &= \quad\$3{,}000 \\
\text{Freight: } 7 \times \frac{41{,}000}{100} &= \quad\$2{,}870 \\
\text{Delivery price} &= \$205{,}870
\end{aligned}$$

The cost of freight to the location under consideration is 7 cents/kg. The weight of the scraper is 40,000 kg, and the weight of the extra attachments is 1000 kg.

$$\begin{aligned}
\text{Less value of four new tires (\$3500 each)} &= \quad\$14{,}000 \\
\text{Delivery price, less tires} &= \$191{,}870
\end{aligned}$$

If the resale or scrap value is considered zero after 5 years and a 40 hr/week, 50 week year is assumed,

$$\text{Hourly depreciation cost} = \frac{191{,}879}{10{,}000} = \$19.18$$

If interest at this particular time is considered annually at 14%, taxes at 3%, insurance and storage at 2% of the average annual investment which is 50% of the initial cost,

$$\text{Hourly cost} = \frac{205{,}870 \times 19 \times 50}{2000 \times 100 \times 100} = \quad\$9.78$$

$$\text{Total hourly ownership cost} = \$28.96$$

If the tire life is estimated to be 3500 hr,

$$\text{Tire replacement cost is } \frac{14{,}000}{3500} = \quad\$4.00$$

$$\text{Tire repairs (15\% tire replacement cost)} = \$0.60$$

General repairs are based on a depreciation period of 10,000 hr and are estimated as follows:

Hourly repair costs = repair factor %

$$\times \text{ hourly depreciation cost} \times \frac{\text{depreciation period}}{10{,}000}$$

Therefore hourly repair costs are

$$\frac{50}{100} \times 19.18 \times \frac{10{,}000}{10{,}000} = \$9.59$$

The repair factor varies from about 40 to 60% for scrapers depending on the job conditions.

$$\text{Fuel cost} = \text{fuel consumption/hr} \times \text{cost/liter}$$
$$= 50 \text{ liters} \times 0.30 \qquad\qquad \$15.00$$

The service cost includes oil, grease, and labor and may be assumed to be one third of the fuel cost for a medium duty cycle.

$$\text{Service cost} = \tfrac{1}{3} \times 15 = \$5.00$$

$$\text{Cutting edge cost} = \frac{\$65 \times 4}{1000} = \$0.26$$

Operator including fringe benefits at $16.00/hr

Operator's cost	= $16.00
Total hourly operating cost	= $50.45
Total hourly ownership and operating cost	= $79.41

An allowance, generally a percentage, for profit, overheads, and supervision must be added to the cost, and this amount depends on various factors. The possible variation and degree of accuracy of the cost per hour are considered as well as the present cash flow of the company and the relative importance of successfully winning the bid for the project. This relative importance is influenced by variable monthly situations. A different markup percentage would be allotted if the company had recently been awarded several contracts compared with a situation where the company had a considerable number of personnel and machines idle because it had not been successful in recent bids for projects.

Since the earth moving item is usually measured in cubic meters, by calculating the production of the machine in cubic meters per hour and knowing the hourly ownership and operating cost, the cost per cubic meter may be entered in the estimate, including an allowance for the profit, overheads, and supervision.

10.7 DEPRECIATION

There are several methods of calculating equipment depreciation. Three methods that are frequently used and approved by the tax authorities are the straight line method, the declining balance method, and the sum of the years digits method.

The straight line method assumes a straight line graph for the value of the machine from the new machine price to zero or the scrap value over the life expectancy of the machine. This method was used in Example 10.1 and is the most common.

Example 10.2

A machine is bought for \$100,000. The life expectancy of the machine is 5 years. Determine the value of the machine at the end of each year assuming a straight line method of depreciation. The machine may be assumed to have no salvage value at the end of year 5.

Answer 10.2

$$\text{Depreciation} = \tfrac{1}{5} \times 100 = 20\%$$

Values are given in Table 10.1

TABLE 10.1   The Straight Line Method of Depreciation

Year end	Depreciation (\$)	Book value (\$)
New		100,000
1	20,000	80,000
2	20,000	60,000
3	20,000	40,000
4	20,000	20,000
5	20,000	0

The declining balance method is sometimes used, which slightly more realistically assumes that depreciation is higher for the first few years of a machine's life. Generally a rate of double the average annual depreciation rate, determined in the straight line method, is multiplied by the value or book value of the machine to give the depreciation for a particular year.

Example 10.3

A machine is bought for $100,000. The life expectancy of the machine is 5 years. Determine the value of the machine at the end of each year using the declining balance method of depreciation.

Answer 10.3

$$
\begin{aligned}
\text{Depreciation rate for each year} &= \tfrac{1}{5} \times 100 \times 2 &&= 40\% \\
\text{Depreciation at end of year 1 (\$)} &= 100{,}000 \times 40\% &&= 40{,}000 \\
\text{Machine value at end of year 1 (\$)} &= 100{,}000 - 40{,}000 &&= 60{,}000 \\
\text{Depreciation at end of year 2 (\$)} &= 60{,}000 \times 40\% &&= 24{,}000 \\
\text{Machine value at end of year 2 (\$)} &= 60{,}000 - 24{,}000 &&= 36{,}000
\end{aligned}
$$

Values are given in Table 10.2.

TABLE 10.2 The Declining Balance Method of Depreciation

Year end	Depreciation ($)	Book value ($)
New		100,000
1	40,000	60,000
2	24,000	36,000
3	14,400	21,600
4	8,640	12,960
5	5,184	7,776

A more sophisticated but not necessarily more accurate method makes allowance for the scrap value or salvage value of a machine at the end of its useful life. The rate of depreciation is calculated from a formula that includes the life of the machine, the purchase price, and the salvage value. The depreciation rate is evaluated and arranged so that the value of the machine is the salvage value at the estimated end of the life of the machine. The usual method is to ensure that the book value of the machine does not fall below the estimated minimum salvage value; for instance, the machine represented in Table 10.2 might have an estimated salvage value of $8500 at the end of the fifth year. This would obviously mean that its value was $8500 and not $7776, and the depreciation during the fifth year is $4460 ($12960, − $8500) and not $5184 as shown in Table 10.2.

For equipment with short lives the declining balance method is not a good method because it tends to overestimate the amount of depreciation. All methods of determining the amount of annual depreciation contain some degree of inaccuracy. There are several factors that can suddenly reduce the value of a machine. However, it is necessary to assign a book value to each machine annually for tax purposes.

In the sum of the years digits method all the numerical values, the digits, of the years to the end of the machine's life are added together. The depreciation is calculated from a factor that represents the reverse order of the numerical values of the year divided by the sum of the years.

Example 10.4

A machine is bought for $100,000. The life expectancy of the machine is 5 years. Determine the value of the machine at the end of each year assuming a sum of the years digits method. The machine may be assumed to have no salvage value at the end of year 5.

Answer 10.4

$$\Sigma_{years} = 1 + 2 + 3 + 4 + 5 \qquad = 15$$

Depreciation factor after year 1 ($) $= \frac{5}{15}$

Depreciation after year 1 ($) $\qquad = \frac{5}{15} \times 100,000$

$$= 33,333$$

Machine value at end of year 1 ($) $= 66,667$

Depreciation factor after year 2 ($) $= \frac{4}{15} \times 100,000$

$$= 26,667$$

Values are given in Table 10.3.

TABLE 10.3   Sum of the Year Digits Method of Depreciation

Year end	Depreciation factor	Depreciation ($)	Book value ($)
New			100,000
1	5/15	33,333	66,667
2	4/15	26,667	40,000
3	3/15	20,000	20,000
4	2/15	13,333	6,667
5	1/15	6,667	0

If a salvage value had been assumed at the end of year 5, a different depreciation could have easily been evaluated. For instance, the depreciation at the end of year 2 would be $\frac{4}{15}$ multiplied by $100,000 less the salvage value.

The loss in value of the machine less any tax concession is the total net loss on a machine. The amount of annual depreciation obtained by any method in this section may only represent an arbitrary amount for tax purposes. The actual secondhand value of the machine from one year to the next may differ from the depreciation value returned for tax purposes. Generally a company has to pay a tax on the profits. If the company makes a loss in some areas or equipment depreciates, an allowance is made for this negative profit in the form of a tax credit, and therefore the company pays less tax.

Machines depreciate according to the amount of usage. Contractors almost always calculate depreciation in terms of time, but this is not accurate if a machine is underutilized. However, the annual costs of keeping an underutilized machine are significant and therefore all machines should be fully utilized if at all possible. In some latitudes the seasonal weather effects can prevent machines being used throughout the year on civil engineering sites.

10.8   EQUIPMENT REPLACEMENT

Complete and accurate records of costs broken down into sections should be kept. The breakdown of these costs are listed under operational costs and ownership costs given in Sections 10.4 to 10.6. These records not only provide valuable information to management for bidding purposes but also are vital in determining the optimum time to replace a machine. The site record of machines should show the repair costs for materials, parts, and labor. When a major repair is necessary or appears imminent, or an old machine has been operating on a project that has been completed, the management can decide at this point, if the repair costs have been accurate and well documented, whether the machine should have a major overhaul or be replaced.

10.9   WORKED EXAMPLES

Note that Examples 10.1 to 10.4 have been inserted at specific points within the text, not at the end of the chapter, to facilitate the learning process and maintain continuity.

Example 10.5

Determine the total hourly ownership and operating costs of the following crawler tractor:

Purchase price	= $180,000
Extras	= $30,000
Shipping charge	= $0.10/kg
Weight of tractor, including extras	= 50,000 kg
Depreciation period	= 10,000 hr
Interest	= 14% of average annual investment
Taxes	= 3% of average annual investment
Storage	= 3% of average annual investment
General repairs	= 60% of hourly depreciation cost
Estimated fuel consumption	= 60 liters/hr
Fuel	= $0.30/liter
Service	
Parts	= 9% of hourly depreciation cost
Greasing, oiling, labor, and so on	= $14/hr
Mechanics, assume 3% time	= $17/hr
Operator, including fringe benefits	= $15/hr

Answer 10.5

Crawler tractor—hourly cost estimate	$	$/hr
Purchase price	180,000	
Extras[a]	30,000	
Freight[b]	5,000	
Delivered price	$215,000	
Hourly depreciation[c]		21.50
Interest, taxes, insurance, and storage		10.75
Total hourly ownership cost[d]		32.25
General repair[e]		12.90
Fuel cost[f]		18.00
Service cost[g]		3.01
Operator[h]		15.00
Total hourly operating cost		48.91
Total hourly ownership and operating cost[i]		81.16

*a*Extras include canopy, counterweight, other tracks, another shaped blade.

*b*Freight costs, say at $0.10/kg. The weight of the tractor including extras is 50,000 kg. Therefore freight costs are $5,000.

*c*Assume depreciation period $40 \times 50 \times 5 = 10,000$ hr. Therefore hourly depreciation $= 215,000/10,000 = \$21.50$ hr.

*d*Assume interest 14%, taxes 3%, insurance and storage 3% of average yearly investment. Assume 2000 hrs/year; that is,

$$\frac{20}{100} \times 215,000 \times \frac{50}{100} \times \frac{1}{2000} = \$10.75/\text{hr}$$

*e*Assume 60% of hourly depreciation cost over a depreciation period of 10,000 hr. Therefore the cost of

$$\text{general repairs} = \frac{60}{100} \times 21.50 \times \frac{10,000}{10,000} = \$12.90/\text{hr}$$

*f*Estimated fuel consumption 60 liters/hr at $0.30/liter; that is, $60 \times 0.3 = \$18.00/\text{hr}$

*g*Service costs:

Parts—assume 9% of hourly depreciation $0.09 \times \$21.50 = 1.94$
Greasing, oiling, labor at $14/hr, $0.04 \times \$14.00 = 0.56$
  4% of time
Mechanics labor at $17, 3% of time $0.03 \times \$17.00 = \underline{0.51}$
                                              $\$3.01/\text{hr}$

*h*Operator including fringe benefits $15/hr.

*i*This figure excludes profit, overheads, and supervision.

Example 10.6

A tractor with blade and ripper costs $250,000. Plot a graph of book value against time using three different methods of calculating the depreciation for the machine assuming a 5 year life. Assume that the salvage value is $20,000 at the end of the fifth year.

Answer 10.6

(a) By the double declining balance method

Double percentage depreciation $= \dfrac{100}{5} \times 2 = 40\%$ for 5 year life

Depreciation at end of year 1 = 250,000 × 40% = $100,000
Depreciation at end of year 2 = 150,000 × 40% = $60,000
Depreciation at end of year 3 = 90,000 × 40% = $36,000
Depreciation at end of year 4 = 54,000 × 40% = $21,600
Depreciation at end of year 5 = 32,400 × 40% = $12,960

Year end	Depreciation ($)	Book value ($)
New		250,000
1	100,000	150,000
2	60,000	90,000
3	36,000	54,000
4	21,600	32,400
5	12,960	19,440
		(20,000)

(b) By the sum of the years digits

Depreciation after year 5 $= \frac{1}{15} \times 230,000 = \$15,333$
Depreciation after year 4 $= \frac{2}{15} \times 230,000 = \$30,667$
Depreciation after year 3 $= \frac{3}{15} \times 230,000 = \$46,000$
Depreciation after year 2 $= \frac{4}{15} \times 230,000 = \$61,333$
Depreciation after year 1 $= \frac{5}{15} \times 230,000 = \$76,667$

Year end	Depreciation ($)	Book value ($)
New		250,000
1	76,667	173,333
2	61,333	112,000
3	46,000	66,000
4	30,667	35,333
5	15,333	20,000

(c)   By the straight line method

$$\text{Depreciation each year} = \frac{230,000}{5} = \$46,000$$

Year end	Book value ($)
New	250,000
1	204,000
2	158,000
3	112,000
4	66,000
5	20,000

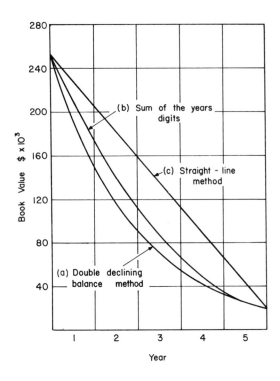

CHAPTER 11

CONCRETE AND FORMWORK

11.1 FORMWORK REQUIREMENTS

When fresh concrete is poured on site, it must be contained in a structure until the concrete has set. This structure, called formwork, maintains the shape and dimensions of the fresh concrete. The formwork usually remains in position until the concrete has attained sufficient strength to support itself. It includes the system of support for the concrete, namely, the surface in contact with the concrete face, and the members that directly support the contact facing material.

Formwork should be strong enough to resist the pressure of the freshly placed concrete. A method of determining the concrete pressures is given in Section 11.2. In addition to the pressures exerted by the concrete, the formwork must resist any other superimposed load that might act on it. A live load of 1.5 to 3.5 kN/m^2 should be used to allow for the normal vertical loads from concreting operations on formwork in a horizontal or near horizontal plane. Wind loads should be considered and these values should be obtained from the relevant national standard on wind loading. If the element of formwork that is being designed is not a major structural component, a wind pressure of 0.32 kN/m^2 at zero height to a pressure of 0.54 kN/m^2 at 20 m height may be assumed in nonsevere exposures. In more severe exposures, such as at the sea coast or estuaries, or at altitudes over 150 m above sea level, a wind pressure exerted on the formwork of 0.54 kN/m^2 at zero height to 1 kN/m^2 at a height of 20 m should be assumed.

To reduce costs formwork should be used as many times as possible. The use of standard modular sizes reduces costs significantly. The modular sizes are usually based on the size of the plywood available; for example, several panels 2.4 × 1.2 m may be fastened together to form larger areas in multiples of 2.4 × 1.2 m or 1.2 × 1.2 m, as shown in Figure 11.1. Plywood is made from an odd number of thin sheets of

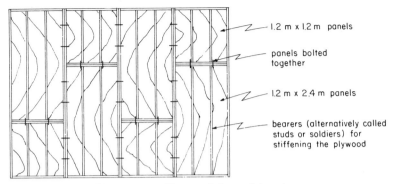

Figure 11.1 Formwork in modular sizes.

timber glued together with the grain of each slice perpendicular to the adjacent sheets; the grains of the two outside slices therefore are parallel. Common thicknesses for plywood on construction sites are 19, 15, and 12 mm.

Bearers or studs are screwed or nailed to the plywood for stiffening. The spacing of the bearers must be based on the greatest possible concrete pressures when standard modular panels are used. Bearers are often made from 150 × 75 mm or 50 mm timber and are spaced at the appropriate centers, which for a 19 mm thick plywood sheet might be about 350 mm, depending on the concrete pressure. Twin wales are then used in conjunction with ties and props to align the formwork. Figure 11.2 shows the positioning of the bearers and wales. The wales are frequently made from 225 × 75 mm timber or specially prefabricated steel members.

To increase the life of the plywood and thus increase its number of uses, a coat of resin may be painted on the plywood faces. This film makes the plywood highly impervious to moisture and prevents any variation in absorbency which creates the grain patterns on the finished concrete face. Textured or plain glass fiber reinforced plastic surfaces can be supplied when a very high number of uses and a high quality finish are required.

Steel is sometimes used for formwork when very many uses are required. Steel replaces plywood as formwork for concreting operations such as the casting of prestressed precast concrete beams, and steel channels or I beams often replace the 225 × 75 mm wales or soldiers.

Glass fiber reinforced plastic forms are used especially for curved surfaces. They are used extensively for dome pans in concrete floors and roofs and have been used effectively for circular column formwork.

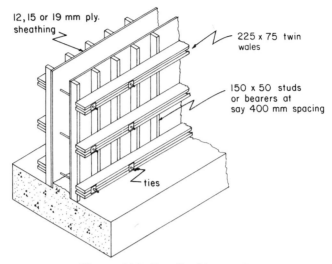

Figure 11.2 Details of formwork.

Additional stiffening can be provided by ribs made of plastics, metal, or timber if required. The plastic surface can produce an excellent concrete surface.

Although the formwork might be able to resist all the pressures exerted on it as far as bending and shear are concerned, the deflection should also be checked. The deflection may be limited by tolerances specified for the permanent concrete structure, but in any case the deflection of members should be limited to the lesser of 3 mm or 0.003 of the span. The position of the structure is also important for deflection. A wavy surface in a concrete member produced by the concrete pressure on the plywood between studs would be undesirable at eye level in a public place but might be acceptable in a tunnel used for services only.

As the loads on formwork are usually only of a temporary nature, the permissible stresses specified may be higher than those for permanent timber structures.

The typical properties of timber used for studs, bearers, soldiers, and wales are shown in Table 11.1. The properties vary depending on the timber species. The permissible stresses vary according to the time duration of the load. For temporary works green timber should be assumed.

Obviously with all these variables the formwork design engineer should ensure that the stresses are appropriate for the type of timber

TABLE 11.1   Properties of Timber for Temporary Formwork

Bending	Safe working stresses (N/mm²)		Modulus of elasticity (kN/mm²)
	Bearing perpendicular to grain	Shear parallel to grain	
9	2	0.9	8

TABLE 11.2   Properties of Plywood Used for Formwork

Thickness (mm)	Working stresses (N/mm²)			E (kN/mm²), parallel to face grain
	Bending along face grain	Bending perpendicular to face grain	Shear parallel to face grain	
12	11	7.6	0.9	8.5
15	10.5	7.8	1.0	7.3
18	10	7.8	1.1	6.9

used and the expected load duration. The typical properties of plywood are shown in Table 11.2.

11.2 CONCRETE PRESSURES

Excessive deformation of formwork arising from the pressure of concrete can result in extremely expensive remedial work. The unit cost of formwork in relation to the overall cost of a concrete structure can be high, so it is extremely important for the formwork to be designed with an accurate knowledge of the magnitude of the concrete pressure.

If concrete behaved as a true fluid, the pressure would be the same in all directions and would be equal to the density of the concrete multiplied by the depth. Fresh concrete tends to behave like a liquid when it is being vibrated, but it generally does not display an equivalent triangular hydrostatic pressure shape because the concrete stiffens at a certain depth in the formwork and the pressure decreases to zero. This effect is shown diagrammatically in Figure 11.3a. Therefore stiffening of the concrete causes a maximum pressure to exist at a particular

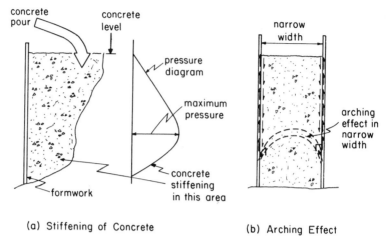

(a) Stiffening of Concrete   (b) Arching Effect

Figure 11.3   Reduction in concrete pressure.

time. This time is dependent on the rate of chemical reaction, hence the temperature, and the workability of the concrete. As the stiffening process proceeds, the concrete begins to develop a shear strength, and at a certain depth it can support a layer of fresh concrete above, without an increase in lateral pressure. The stiffening process also depends on the mode of vibration.

There is also a reduction in the lateral pressure of freshly placed concrete in a wall or similar structure where the section is narrow, that is, less than 500 mm. Frictional forces develop between the concrete and the face of the formwork, see Figure 11.3b, producing an arching effect. These frictional forces are particularly important in narrow sections, where the surcharge volume is comparatively small, and thus create an arching effect that consequently reduces the lateral pressure.

The impact effect on freshly placed concrete when concrete is discharged from a height increases the lateral pressure. CIRIA Research Report No. 1[20] states that the effect of impact is not very significant because in shallow lifts impact surcharges are small and in high lifts the total pressure including impact is rarely greater than the later pressure to which the formwork will be subjected. When concrete is discharged from a height greater than 2 m, an extra 10 kN/m² pressure should be allowed for impact.

The maximum pressure P_{max} of fresh concrete cannot be greater than the total head of concrete. That is,

$$P_{max} \not> \Delta H \tag{11.1}$$

where Δ is the density of concrete in kN/m³, equal to 23.6 kN/m³, and H is the height of the concrete pour.

In thin concrete sections, where the arching effect can limit the concrete pressure, the pressure can be determined from the following formula:

$$P_{max} = \left(3R + \frac{d}{10} + 15\right) \text{kN/m}^2 \qquad (11.2)$$

where d is the thickness of the section in mm and R is the rate of placing in m/hr. This formula is not valid if the thickness of the concrete section exceeds 500 mm.

Because of the stiffening of the concrete in the lower depths of the pour, a triangular pressure diagram should not be assumed, as mentioned previously. The formula that accounts for the stiffening effect is as follows:

$$P_{max} = (\Delta RK + 5)\text{kN/m}^2 \qquad (11.3)$$

where K is a correction factor that accounts for the workability of the concrete and the concrete temperature, expressed in hours. Refer to Figure 11.4.

In the lower part of a concrete pour the concrete stiffens during slow pour rates. During high pour rates the poker vibrators progress upward, thus reducing the pressure at lower depths. For these reasons an arbitrary ceiling of 150 kN/m² is fixed for a maximum possible pressure.

As the arching and stiffening effects reduce the concrete pressures, the least value obtained from the three equations may be taken as the maximum concrete pressure on the formwork.

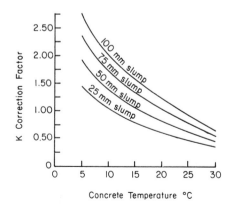

Figure 11.4 Correction factor versus concrete temperature.

Example 11.1

Calculate the maximum pressure generated during a concrete pour for a retaining wall 5.0 m high and 300 mm wide. The concrete$_{mix}$ is designed to have a slump of 50 mm and it is expected that the rate of pour will be 3.5 m/hr. Assume that the concrete temperature is 15°C and that there is continuity in vibration. Assume that the concrete may be discharged from a height greater than 2.0 m.

Answer 11.1

(a) P_{max} is calculated by use of formulas:

1. By height

$$P_{max} = \Delta H \qquad \text{(see equation 11.1)}$$
$$= 23.6 \times 5$$
$$= 118 \text{ kN/m}^2$$

2. Arching effect

$$P_{max} = 3R + \frac{d}{10} + 15 \qquad \text{(see equation 11.2)}$$

$$= 3 \times 3.5 + \frac{300}{10} + 15$$

$$= 55.5 \text{ kN/m}^2$$

3. Stiffening effect

$$P_{max} = (\Delta RK + 5) \qquad \text{(see equation 11.3)}$$
$$= (23.6 \times 3.5 \times 1.1) + 5$$
$$= 95.8 \text{ kN/m}^2$$

Hence

$$P_{max} = 55.5 + 10 = 65.5 \text{ kN/m}^2$$

including 10 kN/m² for impact.

An alternative method is to use the P_{max} design diagram[20] shown in Figure 11.5.

(b) P_{max} is calculated by use of the design diagram. The pressure is measured radially from 0, where 1 cm is equivalent to 10 kN/m², along the rate of pour line to intercept the curve for concrete temperature or arching limit, whichever is less. That is, from measurements

$$P = 5.5 \text{ cm at } 3.5 \text{ m/hr with } d = 300 \text{ mm}$$

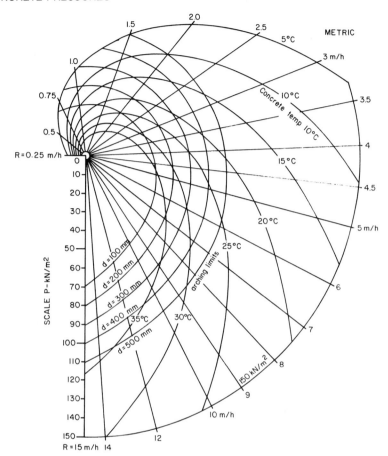

Figure 11.5 P_{max} design diagram.

The chart does not include the 10 kN/m² surcharge allowance for impact; therefore

$$P_{max} = (5.5 \times 10) + 10$$
$$= 65 \text{ kN/m}^2$$

Although equations 11.1 to 11.3 and Figure 11.5 enable the young construction engineer to grasp the significances of various criteria considered in formwork design, they have now been superseded, and the limiting values for concrete pressure on formwork, to the nearest 5 kN/m², can be obtained from the following tables. The tables[21] are only applicable to Portland cement concretes without cement replacement materials or admixtures, and to internally vibrated concrete. The design pressure is again taken as the least numerical value obtained, and an allowance for impact arising from placing operations is not normally necessary.

The hydrostatic pressure, 25 kN/m² per meter of height, is taken from Table 11.3.

When the concrete width is less than 500 mm and an arching effect can reduce the pressure, it is taken from Table 11.4.

The concrete pressure reduction due to the stiffening limit is taken from Table 11.5.

The American Concrete Institute[22] suggests the following formula for calculating the maximum lateral pressure on formwork P_i for prescribed conditions of temperature, rate of placement, vibration, weight of concrete, and slump value, for walls with R_i not exceeding 7 ft/hr:

$$P_i = 150 + \frac{9000\,R_i}{T_i} \qquad (11.4)$$

with a maximum of 2000 psf or 150 h_i, whichever is less. For walls with R_i greater than 7 ft/hr

$$P_i = 150 + \frac{43{,}400}{T_i} + \frac{2800\,R_i}{T_i} \qquad (11.5)$$

TABLE 11.3 Concrete Pressure—Height Limit

H (m)	1	2	3	4	5	≥6
P (kN/m²)	25	50	75	100	125	150

TABLE 11.4 Concrete Pressure—Arching Limit

	R (m/hr)											
d (mm)	1	2	3	4	5	6	8	10	15	20	30	≥40
150	35	35	40	45	45	50	55	60	75	90	120	150
200	40	40	45	50	50	55	60	65	80	95	125	150
300	50	50	55	60	60	65	70	75	90	105	135	150
400	60	60	65	70	70	75	80	85	100	115	145	150
500	70	70	75	80	80	85	90	95	110	125	150	150

Concrete pressure (kN/m²)

TABLE 11.5 Concrete Pressure—Stiffening Limit

Slump (mm)	Concrete temperature (°C)	R (m/hr)									
		1	1.5	2	2.5	3	4	5	6	7	≥8
50	5	50	70	95	115	135	150	150	150	150	150
	10	40	55	70	85	100	135	150	150	150	150
	15	40	45	55	65	75	100	125	150	150	150
	20	35	40	45	50	55	70	90	105	125	150
75	5	60	85	110	140	150	150	150	150	150	150
	10	50	65	85	105	125	150	150	150	150	150
	15	40	50	65	80	95	125	150	150	150	150
	20	35	40	50	60	70	90	115	135	150	150
100 to 150	5	70	100	130	150	150	150	150	150	150	150
	10	55	75	100	120	150	150	150	150	150	150
	15	45	60	75	90	110	150	150	150	150	150
	20	35	45	55	70	80	110	130	150	150	150
		Concrete pressure (kN/m²)									

(Reference 21)

with a maximum of 2000 psf or 150 h_i, whichever is less, where, for equations 11.4 and 11.5 only,

P_i = maximum lateral pressure (psf)
R_i = rate of placement (ft/hr)
T_i = temperature of concrete in the form (°F)
h_i = maximum height of fresh concrete in the form (ft)

Here i denotes imperial units.

11.3 FORMWORK DETAILS

The design, detailing, and erection of formwork should be supervised by experienced site personnel. Errors are sometimes made when the method and ease of dismantling or striking the formwork are not considered. Protruding reinforcement, necessary for the overall structural integrity and used in subsequent concrete pours, should be considered when necessary. It is important for the young engineer to discuss his proposals of formwork detailing with the foreman carpenter. The foreman carpenter's observations, based on years of experience, should enable the less experienced formwork designer to foresee any impracticalities in the formwork proposals.

Section 11.2 shows that the rate of pour, temperature, method of placing, extent of vibration, concrete mix proportions, and water-cement ratio contribute to the amount of pressure on the formwork. Any restrictions on the amount of pressure on the formwork resulting from the assumptions made in design should be clearly stated on the drawings. For example, it might be necessary to give the maximum rate of pour on a drawing. When prefabricated panels are made for multiple uses, an attempt should be made to accurately predict the worst pressure that the panel will be subjected to and design the panel accordingly.

The surface finish of concrete is largely influenced by the type and quality of the formwork. Various finishes are graded and specified. A surface where joints or small blemishes are seen may be permitted. Another type of finish, with high quality concrete and only very minor blemishes with no staining or discoloration, may be specified. Yet another finish may call for a high quality concrete but minor imperfections that can be remedied afterward by using a cement–fines aggregate paste. Obviously the type of finish selected depends on the position of the element and how frequently and easily it can be seen. The lower stem of a reinforced concrete abutment, for instance, requires a concrete of consistent and predictable quality but, as the front and back will be covered by fill, it is futile to spend a lot of time and money on the quality of the surface finish.

Sometimes concrete, called blinding or the mud slab, of about 75 mm thickness is laid for the construction of reinforced concrete bases so that in soft soil conditions the excavated area in the vicinity of the base can remain clean and dry. If forms have already been prepared on site in modular lengths, great care should always be taken to prevent the modular lengths from being sawn through and shortened to suit a short dimension on site. Waste can often fairly easily be prevented by a method similar to the one shown in Figure 11.6, where the formwork for a typical base is shown.

To help position the column or wall an upstand of concrete, called the kicker, of about 75 mm depth is often cast at the same time as the base. There are alternatives to the method shown in Figure 11.6, such as concreting the kicker after the base, inserting a key within the base, or

Formwork for Bases

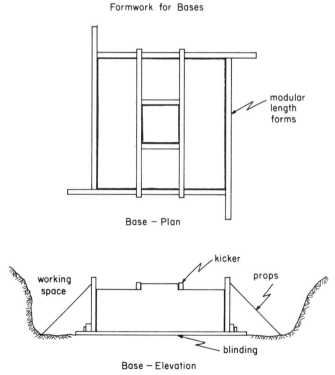

Figure 11.6 Formwork for bases.

using steel pins to position the wall formwork. The bottom of the stem of an abutment or retaining wall is an extremely important area where the highest bending moments and forces may be present. This is sometimes overlooked on site. The concrete in this area should be particularly well vibrated and care should be taken to ensure that the concrete has attained the specified strength. The kicker cast in situ with the base is therefore the best solution technically, as this ensures a high quality concrete of sufficient strength. Practically, however, it has a drawback. The timber used to form the kicker is often in a convenient place for the concreting gang to stand on, and during the concreting operation the timber can become displaced, necessitating remedial work to erect the wall formwork in the exact position. This problem highlights the importance of making robust formwork that is capable of withstanding not only maximum concrete pressures but also the forces that are present during the concreting operations. The significance of the required accuracy in the positioning of formwork is therefore important.

Formwork for walls is propped by some means to prevent it from tilting. The ties shown in Figure 11.7 prevent the two faces of the formwork from moving away from each other. The reinforcement cage, not shown on the diagrams for reasons of clarity, and small concrete or plastic spacer blocks, which ensure the correct amount of cover be-

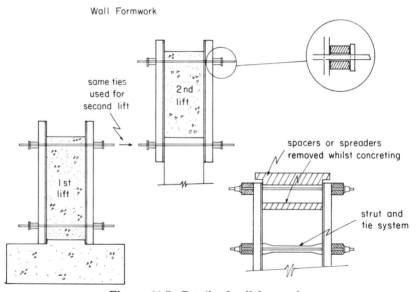

Figure 11.7 Details of wall formwork.

tween the formwork and the reinforcement, prevent the two faces from moving toward each other. Inward movement is also prevented by the use of temporary timber struts or a strut system incorporated within the ties, as shown in Figure 11.7.

When large areas of vertical formwork are required, it is convenient to prefabricate large standard panel systems made from 1.2 × 2.4 m sheets of plywood. The maximum concrete pressures must be considered in design for the spacing of the bearers. The rate of stiffening, for instance, is much slower at lower temperatures. Gusts of winds can make the handling of such panels difficult. It is therefore necessary to have a good system of handling the panels similar to the method shown in Figure 11.8.

Formwork for the soffit or undersurface of roofs, floors, or bridge decks may consist of plywood and ribs of 150 × 75 mm resting on 225 × 75 mm longitudinal timbers which are supported on a braced grid of conventional scaffolding or a proprietary unit frame system. The exact

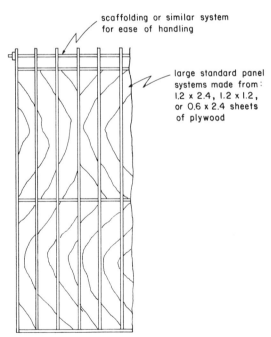

Figure 11.8 Large standard panel systems.

Soffit

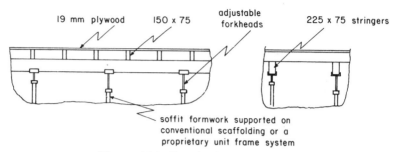

Figure 11.9 Details of soffit formwork.

level of the longitudinal members, which should take into account any forces that create deflection, is obtained by the adjustable fork-heads as shown in Figure 11.9.

Ties usually are 50 mm less than the wall thickness. The she bolt is recovered for many uses, and the holes left in the concrete surface are filled in unless they are to be used again for subsequent concrete pours. Snap ties, in which the projecting part of the tie is disposable, are occasionally used. Refer to Figure 11.10.

The spacing of the ties may be deduced from the values of the maximum safe working loads of the ties given in Table 11.6.

11.4 DESIGN OF FORMWORK

Formwork should be designed to withstand all the vertical loads and lateral pressures to which it might be subjected. The formwork should

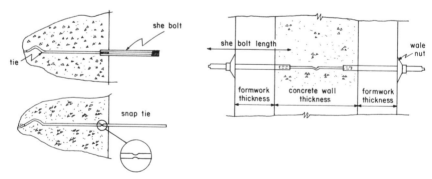

Figure 11.10 Details of ties.

TABLE 11.6   Safe Working Load of Ties

She bolt diameter (mm)	Maximum safe working load (kN)
14.3	33
17.5	57
22.2	89

maintain its shape, alignment, dimensions, and level. Working stresses should be used in design. Use of the principles of limit state design is considered inappropriate, as formwork design, although using the same structural principles as the design of permanent works, is different. The temporary nature of the loading is utilized. Damage to formwork on site after several uses must be considered. Handling and exposure to the weather elements may also damage formwork and limit the numbers of uses.

The pressure exerted by freshly placed concrete should be calculated using the information given in Section 11.2. Self-weight of formwork and concrete should be considered and allowed for when necessary. A general imposed loading of 1.5 to 3.5 kN/m^2 on soffit formwork should be allowed for. This general imposed loading includes an allowance for the load caused by construction workers, minor plant and equipment such as vibrators, stacking of materials, and the surchange and impact of concrete, which are significant for thin slabs.

The stresses of typical properties of timber used for formwork are given in Table 11.1. These values are approximate and are based on various species used for formwork. An alternative to giving the properties of timber according to species is to stress grade timber. The British Standard CP112 (1972) Amdt. No. 1[23] gives the values given in Table 11.7 for any timber that conforms to the SS (Standard Specification) grade. This is the grade of timber in general use for formwork.

The values given here assume a moisture content up to about 28%, which satisfies the condition of most timber used on site. With increasing moisture content timber loses strength. Stresses up to about 40% higher may be assumed if the timber is used in a dry location.

Various grades are marked on timber. The values of the stresses given in Table 11.7 may be modified by multiplying by a factor of 1.5 if the duration of loading only acts for one day and by a factor of 1.3 if it acts for one month. This is because the mechanical properties of timber, except for the modulus of elasticity, E, are time dependent. A

TABLE 11.7   Properties of Stress Graded Timber

	SS grade stresses (N/mm²)			Modulus of elasticity (kN/mm²)	
Bending	Bearing perpendicular to grain	Shear parallel to grain		E_{min}	E_{mean}
5.9	1.0	0.7		4.7	8.2

modification factor should not be applied to the modulus of elasticity, although if the load is shared by two members a value for E of 5.7 kN/mm² may be assumed. If the system contains four or more members, the mean value of E may be assumed.

For the design of the plywood sheeting and bearers the maximum bending moment that can occur, assuming that the beam or plywood is continuous over three or more spans, is $0.107wl^2$ kN-m when subjected to a uniform load. The maximum shear force that can occur is $0.61wl$ kN. These maximum values occur at the center line of the penultimate support. The generally accepted values in formwork design are $0.10wl^2$ for the maximum bending moment and $0.60wl$ for the maximum shear force. If the local reduction in bending moment and shear force arising from the finite thickness of the supports is considered, it can be seen that the lower simplified design values are admissible.

The moment of resistance M_R of a timber beam is

$$M_R = \frac{fbh^2}{6} \tag{11.6}$$

where f = allowable flexural stress
        b = width of beam
        h = height of beam

The maximum applied moment M_{max} is

$$M_{max} = 0.10wl^2 \tag{11.7}$$

where w = uniformly distributed pressure
        l = span

By combining equations 11.6 and 11.7 the span can be found:

$$l = \sqrt{\frac{fbh^2}{0.6w}} \tag{11.8}$$

After the spacing of the supports has been determined, shear and deflection checks should be made.

The maximum shear stress of a rectangular timber beam is

$$s_{max} = \frac{3}{2}\frac{V_{max}}{bh} \qquad (11.9)$$

where V_{max} = maximum shear stress.

The maximum shear force may be assumed to be

$$V_{max} = 0.6wl \qquad (11.10)$$

Therefore combining equations 11.9 and 11.10 yields

$$l \not> \frac{1.11bhs_{max}}{w} \qquad (11.11)$$

When timber is subjected to a uniformly distributed concrete pressure, the maximum deflection is given by

$$\delta_{max} = \frac{kwl^4}{EI} \qquad (11.12)$$

where $k = 0.0130$ for both ends simply supported
  $k = 0.0054$ for one end simply supported and the other end fixed
  $k = 0.0026$ for both ends fixed

The ACI[22] recommends $k = 0.0069$ for a beam continuous over three or more spans.

An assessment should be made to establish which value of k the element of formwork under consideration is likely to be closest to for maximum deflection. Calculations relating to deflections are often inexact. The value of the modulus varies depending on the moisture content of the timber. Dry timber tends to be stiffer than wet timber. Therefore in concreting operations the minimum value of the modulus should be assumed in wet conditions.

The value calculated for the anticipated deflection due to bending should not exceed 0.003 of the span or 3 mm, whichever is less. For spans greater than 1.5 m a deflection greater than 3 mm is acceptable to most formwork designers, whereas for plywood some designers limit deflection to 2 mm or less.

Example 11.2

Design a formwork panel 3.6 × 2.4 m for a retaining wall that is to be 400 mm thick. The concrete pour is 4.5 m high and the estimated rate of pour is 1.5 m/hr. The design of the concrete mix indicates a

mean slump of 50 mm and a concrete temperature of 20°C is anticipated. The plywood available for the panels has the following properties:

$$\text{thickness, } h = 19 \text{ mm}$$
$$\text{allowable flexural stress, } f = 10 \text{ N/mm}^2$$
$$\text{maximum shear stress, } s_{max} = 1.1 \text{ N/mm}^2$$
$$\text{modulus, } E = 6.9 \text{ kN/mm}^2$$

Use equations 11.1 to 11.3 to determine the concrete pressures.

It is unnecessary to make the extra allowance of 10 kN/m² for impact
because of the method of pouring the concrete.

Answer 11.2

It is likely that 2.4 × 1.2 × 19 mm plywood sheets and 150 × 50 mm
timber will be readily available on site. Timber with these dimensions
is therefore used in the design of the panels. Refer to Figure 11.11.

The maximum pressure of the concrete exerted on the formwork
panels is the least value of the following:

$$
\begin{aligned}
P_{max} \ \text{(total height)} \quad &= \Delta H \\
&= 23.6 \times 4.5 \\
&= 106.2 \text{ kN/m}^2
\end{aligned}
$$

$$
\begin{aligned}
P_{max} \ \text{(arching limit)} \quad &= 3R + \frac{d}{10} + 15 \\
&= 3 \times 1.5 + \frac{400}{10} + 15 \\
&= 59.5 \text{ kN/m}^2
\end{aligned}
$$

$$
\begin{aligned}
P_{max} \ \text{(stiffening)} \quad &= \Delta RK + 5 \\
&= 23.6 \times 1.5 \times 0.8 + 5 \\
&= 33.3 \text{ kN/m}^2
\end{aligned}
$$

Therefore a maximum pressure of 33.3 kN/M² is assumed in the
design of the panel. The plywood will span from one bearer to the
adjacent bearer. To determine this span l, see Figure 11.11, equation
11.8 should be used as follows:

$$
\begin{aligned}
l &= \sqrt{\frac{fbh^2}{0.6w}} \\
&= \sqrt{\frac{10 \times 10^6}{10^3} \times 1 \times \frac{19^2}{10^6} \times \frac{1}{0.6 \times 33.3}} \quad \text{(kN-m units)} \\
&= 0.425 \text{ m} \\
&= 425 \text{ mm}
\end{aligned}
$$

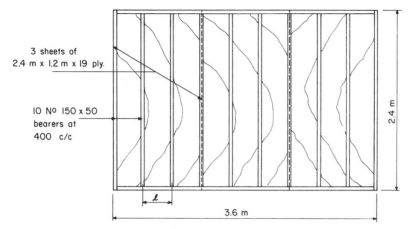

Figure 11.11 A formwork panel.

Loading on 1 m width strip of plywood is considered. As the panel is to be made up from three sheets of 19 mm plywood, the individual sheets will be joined at 1.2 m intervals along the 3.6 m length. Bearers are required at these intervals to stiffen the edges of the plywood and to ensure that there will be no grout leakage or plywood deformation which would spoil the concrete finish. In this situation a suitable spacing for the bearers would normally be 400 or 300 mm.

The 425 mm spacing for the bearers was obtained by using a formula in which the maximum flexural stress of the plywood was considered. Calculations should now be done to verify that the plywood is adequate in shear and deflection. A span of 400 mm is assumed so that join lines of the plywood sheets are stiffened.

The use of equation 11.11 ensures that the maximum permissible shear stress will not be exceeded if

$$l = \frac{1.11 bh s_{\max}}{w}$$

$$= \frac{1.11 \times 1}{33.3} \times \frac{19}{10^3} \times \frac{1.1 \times 10^6}{10^3} \text{(kN-m units)}$$

$$= 0.697 \text{ m}$$
$$= 697 \text{ mm}$$

Hence shear is not critical.

The maximum deflection is given in equation 11.12,

$$\delta_{max} = \frac{kwl^4}{EI}$$

$$\simeq 0.0069 \times 33.3 \times 0.400^4 \times \frac{1}{6.9 \times 10^6} \times \frac{12 \times 10^9}{1 \times 19^3}$$

$$= 0.00149 \text{ m}$$
$$= 1.49 \text{ mm}$$

Thus the value for δ_{max} is less than 2 mm but not less than 0.003 times the span; that is, $0.003 \times 400 = 1.2$ mm. However if the clear span had been used in the equation, that is, $400 - (2 \times 25) = 350$ mm, the value of δ_{max} would have been only 0.87 mm. Therefore the calculations considering shear and deflection do not indicate that the 400 mm span should be altered.

A similar method of design can be used to calculate the spacing of the twin wales. If the maximum safe working load of the ties is 57 kN, one tie is required for every $57/33.3 \text{ m}^2 = 1.7 \text{ m}^2$ of formwork.

The ACI publication "Formwork for Concrete"[22] gives the following formula for P_{max} for walls with R not exceeding 7 ft/hr (2.13 m/hr) (refer to equation 11.4):

$$P_i = 150 + \frac{9000\, R_i}{T_i}$$

with a maximum of 2000 psf or $150h_i$, whichever is less,

where P_i = maximum lateral pressure (psf)
            R_i = rate of pour (ft/hr)
            T_i = temperature of concrete (°F)
            h_i = maximum height of fresh concrete (ft)

Substituting the values of the preceding example, that is, $R = 1.5$ m/hr = 4.92 ft/hr and $T = 20°C = 68°F$, yields the estimated concrete pressure:

$$P_{max} = 150 + \frac{9000\, R_i}{T_i}$$

$$= 150 + \frac{9000 \times 4.92}{68}$$

$$= 801 \text{ psf}$$
$$= 38 \text{ kN/m}^2$$

Note that in addition to flexural, shear, and deflection checks in formwork design bearing stress checks are sometimes necessary.

Bearing stresses should be checked where timber beams lie across each other perpendicularly or around the area of tie washers or similar metal plates which exert a fairly high local pressure.

There are tables available which facilitate formwork design, but the young engineer should understand the fundamental principles before using quicker methods of design. The failure of temporary works contributes to a disproportionate number of accidents on site. Although complete collapse of formwork is rare, partial failure because of excessive formwork deformation is more common than it should be.

11.5 TRAVELING FORMWORK

Vertical slipform systems are suitable for the rapid and economical construction of silos, chimneys, towers, building cores, bridge piers, and any similar tall concrete structures. The construction of tapering cross sections is possible by this method.

The system[24] shown in Figure 11.12 moves upward to form the walls. It is raised by hydraulic jacks controlled and powered from a central control unit. The hydraulic jacks are attached to the formwork system by an adjustable steel frame and are automatically clamped to 50 mm diameter tubes that generally run continuously through the completed structure and relieve the recently laid concrete of the system's weight. Alternatively, the tube or bar can be removed and reused as the concrete pour progresses upward. The jack is attached to the lower of two ties that tie the panels together. The ties and vertical members of the frame are usually made of steel. The frames are positioned around the structure about 2.000 to 2.500 m apart.

The sliding formwork is made of timber with steel infill panels generally about 1.250 m high, with variable lengths depending on the type of structure. The wall thicknesses can vary considerably, with a minimum thickness of 125 mm.

After the reinforcement has been placed and any boxes required for voids or cladding are fixed into position, the concrete is cast and vibrated in layers of between 100 and 200 mm of thickness. When the formwork is full, it is raised very smoothly in successive stages of about 200 mm at a time.

The concrete is poured continuously without interruption. Rates of vertical progress vary depending on the type and simplicity of the structure, but usually average about 200 to 350 mm/hr. The rate of progress is also governed by the atmospheric conditions and type of cement used. The concrete released from the formwork after 5 hr

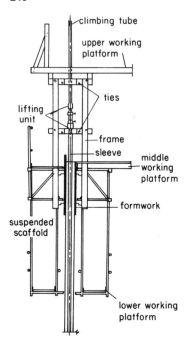

Figure 11.12 Vertical slipform system.

or more is only subject to the compressive weight of about 1.25 m of concrete. Rapid hardening cement is not always necessary, and normal Portland cement has been used. The water/cement ratio is important, and the concrete should be fairly dry when placed. A quality control technique may be utilized in which an accelerated curing system enables tests to be made on the concrete in 24 hr.

Costs are reduced because construction time is greatly reduced after the initial erection procedure, although shift working increases the labor cost.

Small blemishes on the formwork create scratch lines running vertically up the concrete surface which may require surface dressing on the lower platform, although in the case of internal cores the surface provides an excellent key for subsequent finishing work.

Horizontal traveling formwork consists of reusable formwork mounted on mobile frames. When concrete in a section of a structure has cured sufficiently so that the formwork can be released, the frame is rolled along to the next section of the structure. This process can be repeated many times and is particularly suitable for structures that are long and possess a constant cross section such as walls, abutments, roofs, tunnels, culverts, and subways.

Traveling formwork is usually designed and fabricated for each particular job, although sometimes previously used frames and formwork may be adapted to a new type of job. Only long structures make the initial time and effort put into the construction of the traveling formwork system an economical proposition. The horizontal traveling formwork varies in many ways, but a straddled traveler for casting concrete walls is shown in Figure 11.13.

The formwork is mounted on the straddled traveler, which can be made from a steel frame or scaffolding. The complete assembly is mounted on wheels that run either on rails or on a well prepared level surface. The formwork may be released and reerected by the use of winches, jacks, or turnbuckles. Occasionally movement may be mechanized.

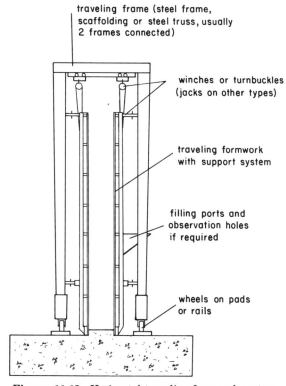

Figure 11.13   Horizontal traveling formwork system.

11.6   TRANSPORTING CONCRETE

Concrete should be transported from the mixer to the formwork as quickly as possible so that the necessary workability can be maintained. The method of transportation and placing should ensure that segregation does not occur. Concrete should be placed and compacted as soon as practicable after mixing. Delays are permitted if the concrete can be effectively placed and compacted without the addition of extra water. Concrete may be allowed to fall freely through any height provided that it does not segregate and the impact does not displace the reinforcement or the formwork.

There are several methods of producing concrete. Mixture of the aggregate, sand, cement, and water may be done at a batching plant on site, the concrete then being transported to its destination. Alternatively, the appropriate amount of materials and water may be loaded into a truck mixer, mixed at the depot, and then transported to site. The addition of water on site is possible but is usually not permitted, particularly for concrete intended for structural use. It is necessary for the truck mixers to agitate the concrete no matter what sequence and method of mixing have been adopted. Normally the concrete must be discharged within 2 hr of the cement coming into contact with the damp aggregate, although it has been shown in some cases that the strength of the concrete is not reduced beyond 2 hr. However, a time comes when workability and compactability deteriorate. Normally the ready mixed concrete supplier expects his 6 m³ truck to be at the point of discharge for a maximum of ¾ hr. For heavily reinforced sections, concrete of a higher workability should be considered to improve the rate of discharge.

The use of ready mixed concrete transported in agitator trucks has become widespread over the last 20 years. This trend will continue as labor rates increase. Because of continuous supply, the ready mixed concrete industry can provide expert knowledge for economical mix designs, and the supply of concrete with consistent properties is usually reliable.

It is extremely difficult to compare the costs of different methods of transportation. Costs for haul roads, haulage to site, temporary works, supply of materials and services, hire rates, or internal company machine costs all need to be considered carefully. The effect of bad weather and supply of skilled labor also must be assessed.

11.7   EQUIPMENT FOR TRANSPORTING CONCRETE

On a construction site concrete must be transported in a horizontal or vertical direction, often in a combination of both directions.

The most common method of moving ready mixed concrete long distances on site or on the public highway is by means of an agitator truck mixer, or ready mix truck as it is popularly known. The volume of concrete that can be carried varies between 3.0 and 10.0 m^3 depending on the type of truck. The concrete is agitated by the slow rotation of the drum while the vehicle is traveling, after all the materials have been mixed together by the rotation of the drum at faster revolutions. The materials for the ready mixed concrete are loaded into the drum through the charge hopper. Inside the drum there are often one long and one short blade together with one large area transverse paddle blade to produce an effective mixing zone in the front of the drum. The concrete is discharged through a discharge chute, which can be swung through 180° horizontally and about 25° vertically, into dumpers or skips or directly into place.

Over the last few years the method of pumping concrete has developed considerably. This three dimensional method of concrete transportation is used in many situations. Both the concrete pump and an articulated hydraulic boom are mounted on one truck, and in one operation receive concrete from an agitator truck and discharge the concrete at the desired place at a different vertical and horizontal location.

The pumps alone can deliver up to about 100 m^3/hr. They can pump concrete as high as 75 m or up to about 350 m horizontally. The delivery system is usually through 100 or 125 mm diameter tubes, and an almost continuous flow of concrete is maintained by the operation of two concrete pistons in sequence. When mounted on trucks the concrete is poured into a hopper at the rear of the truck and is pumped through the pipeline. Articulated booms of 25 m are common and models now exist with booms up to 50 m.

Other methods of transporting concrete on site access roads include the use of dumpers and large open topped skips, of 2 to 3 m^3 volume, which pivot hydraulically and are mounted on trucks. The skip on the dumper is hydraulically operated and can rotate through 180°.

For other methods of horizontal or slightly inclined concrete transportation, belt conveyors or monorail systems may be considered. Belt conveyors may be used when a fairly high rate of pour is required.

Machines are used frequently on construction sites for various operations, not solely for concreting operations. Obviously these machines, such as cranes, should be used when available even if they are not the best system of transporting concrete, provided that the desired rate of production can be maintained.

11.8   REMOVAL OF FORMWORK

After the concrete has obtained a cube strength of 10 N/mm² or twice the stress to which it will be subjected, the formwork may be removed. If test results are not available formwork can be removed in accordance with the time periods given in Table 11.8, which is reproduced from B.S. CP110: Part 1: 1972.[25]

On jobs where the engineer has not made any provision for approval of shore and formwork removal based on strength, ACI recommends[22] that the minimum times before the formwork and supports are removed be based on the times given in Table 11.9.

The period before striking formwork should be amended according to the surface temperature of the concrete or any additives that were added to the concrete. To avoid damage to decorative surface features it may be necessary to increase the times before striking vertical formwork to walls or columns. Great care should be exercised in the design of formwork to prevent internal voids. Some tapering or special narrow closure strips at the edges often facilitate the removal of such formwork. Protruding reinforcement, particularly if it is not straight, can cause problems when striking formwork.

When formwork is removed, a short period after the concrete has been placed, suitable curing methods should follow immediately to insulate and protect the concrete from low or high temperatures.

TABLE 11.8   Minimum Period Before Striking Formwork (B.S. C.P. 110)

Type of formwork	Surface temperature of concrete	
	16°C	7°C
Vertical formwork to columns, walls, and large beams	9 hr	12 hr
Soffit formwork to slabs	4 days	7 days
Props to slabs	11 days	14 days
Soffit formwork to beams	8 days	14 days
Props to beams	15 days	21 days

TABLE 11.9 Minimum Period Before Striking
Formwork (ACI)

	12–24 hr
Walls	12–24 hr
Columns	12–24 hr
Sides of beams and girders	12–24 hr

	Where design load is	
	$<DL$	$>DL$
Floor slabs:		
under 3 m clear span		
between structural supports	4 days[a]	3 days
3 to 6 m clear span	7 days[a]	4 days
over 6 m clear span	10 days[a]	7 days
Joist, beam or girder soffits:		
under 3 m clear span		
between structural supports	7 days[a]	4 days
3 to 6 m clear span	14 days[a]	7 days
over 6 m clear span	21 days[a]	14 days

[a]Where formwork can be removed without disturbing the
shores, half the time may be used if it is not less than 3
days.

Very careful control should be exercised, and approval obtained from
an engineer who has knowledge of the design of the structure, when
soffits are repropped so that the plywood can be reused at an earlier
date. When striking soffit formwork, props in the central area of the
span should be lowered first. The lowering should be done in small
increments in a predetermined sequence, and the operation repeated
for another small incremental movement.

Release agents are applied to formwork faces so that the formwork
may be easily released from the concrete at the time of striking. The
choice of the type of release agent depends on the absorbency and type
of surface finish of the formwork, the type of concrete, and the quality of
finish. One or more applications of the appropriate release agent
should be given immediately prior to the first concrete pour, and one
coat should be applied prior to each subsequent concrete pour.

11.9 ECONOMICS IN FORMWORK

Formwork costs often comprise up to 30 to 60% of the overall cost of
concrete work on a construction job. It is therefore very important that

considerable thought be given to ways of economizing in formwork. In the construction process there are two areas where economies can be made. First, the design engineer can continuously examine the permanent structural design to ensure that it is not creating unnecessary expense in the temporary works. Second, the formwork designer and the foreman carpenter can economize by a critical examination of their methods and plans. In both cases critical assessments are more easily made by the more experienced person.

The main objects in formwork economy are to maximize the number of uses of formwork, to use standard sizes to prevent wasteful cutting, to use the appropriate material for the formwork, and to use proprietary forms if a saving can be made on finishing work.

The engineer designing the permanent concrete structure should remember that frequent changes in dimensions of structural elements should be avoided. Uniform depths, widths, cross sections, and diameters often enable very great savings simply by the number of uses of the formwork. Irregularities in the concrete form should be reduced to a minimum. Some features often seen on concrete structures reduce the amount of concrete used, but the money saved is less than the additional expense to the formwork. This is particularly true for jobs where only a small volume of concrete is required and where ready mixed concrete is used. For instance, a concrete member might require 17.8 m^3 of concrete. A reduction in the volume of concrete to 17.2 m^3 could be made in the design, but more elaborate formwork would be required. But, even disregarding the extra cost of the formwork, the truck volume of concrete ordered would be 18 m^3 in both cases.

The design engineer should attempt to keep beam widths, beam depths, and column dimensions constant for several floors in reinforced concrete multiframed buildings so that the formwork can be reused as much as possible. If several bridge decks with various spans are to be constructed on a highway project and circular reinforced columns are to be used to support the bridge decks, it is more economical for the structural engineer to design all the columns with the same diameter, even though some of the smaller spans may require less concrete according to the design calculations. The multiple use of the same column forms enables the contractor to quote a lower price at the tender stage.

Similar architectural or aesthetic finishes may be obtained at considerably lower formwork costs. If it is felt that a concrete wall would have a more pleasing appearance by breaking the flat surface with horizontal lines, a similar effect may be produced by a protrusion or rustication as shown in Figure 11.14. However, the formwork costs differ

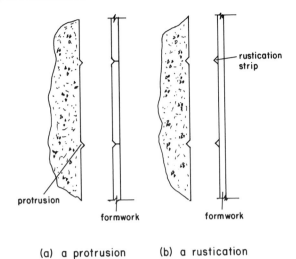

rustication strip

protrusion

formwork   formwork

(a) a protrusion   (b) a rustication

Vertical Cross-section of Wall

Figure 11.14   Features on a concrete wall.

substantially. It is far cheaper to attach the rustication strip to the plywood than to form a concrete protrusion. This considerable saving applies to formwork materials and labor, with little or no increase in concrete for cover requirements, unless prefabricated special purpose forms are used.

The engineer may also assist in formwork economy by not specifying unreasonable qualities of finish and unreasonable accuracy on dimensions.

On several types of contract it is possible to obtain the contractor's view of the proposed structural design before the project has started. Alterations suggested by the contractor may produce very great overall savings, particularly if he can use some multiple use proprietary formwork from a previous project. For this reason the engineer should be prepared to very carefully consider any suggestions made by the contractor even after the job has begun. It is the duty of the engineer to update his knowledge of recent advances in formwork materials and accessories.

The formwork designer should ensure that all the formwork does not exceed the permissible working stresses in bending, shear, and bearing when subjected to the maximum pressures and loads. The design of the formwork should also prevent excessive deflections. Overdesign should be avoided. If only one type of permissible stress is exceeded and the other stresses are well under the permissible stresses, the most economical solution might be to locally reinforce or alter the formwork to reduce the actual stress rather than to completely change the formwork to a more conservative design. After the strength has been verified, the most important economic factor for the formwork is the amount of reuses it is capable of. A cheaper accessory that might damage the formwork or take a longer time to fix should not be used in preference to a more suitable but expensive accessory. Good, clear working drawings for all but the most simple formwork reduce the risk of errors on site. The use of construction joints on large areas of concrete might increase the number of concrete pours but could substantially reduce the amount of formwork required.

The foreman carpenter can reduce site costs for formwork by prefabricating in the workshop, in modular sizes if possible, all the formwork that need not be made on site. The foreman carpenter should also ensure that the formwork is carefully cleaned and oiled after use. Accessories should be well maintained and formwork should be struck as soon as possible for reuse.

11.10 PRACTICE AND SAFETY

There are standard procedures that should be maintained on every construction project to ensure good practice and substantially reduce the possibility of errors and even accidents on site.

The engineer or architect should ensure that there are clear and adequate drawings and specifications of all the structures on a project. The shapes of the structures should be kept as simple as possible. Any limits such as rate of placing, formwork removal times, repropping method and procedure, special sequences in construction, position of construction joints, and the position of service ducts and bolts should be specified very clearly. Samples of special surfaces should be made available to the contractor. The engineer should also inform the contractor of those structures that require formwork drawings for comment and approval. All work should be regularly or continuously inspected.

The contractor on his part should ensure that the specifications and drawings are thoroughly examined at the tender stage so that a fair and accurate price may be inserted in the documents. Any queries should be discussed at the tender stage so that solutions can be worked out for any anomalies that might arise. The rates should be quoted for the specified methods, but any alternative methods favored by the contractor should be discussed with the engineer and, perhaps after

Plate 11.1 Poker vibrators for concrete. Courtesy Wacker Canada Ltd., Mississauga, Ont., Canada.

satisfactory discussions, a rate can be quoted for the preferred method. The contractor is required to ensure that the supervision, materials, and workmanship maintain a high quality, and that all the standards and specifications are met so that a safe and pleasant appearance is achieved as well as a profit on the job.

A considerable proportion of the accidents that occur on site are caused by the failure of the temporary works. A complacent attitude should not and must not be taken in view of the number of accidents and incidences of formwork or scaffolding failures. A good practice, as outlined in this section, is of considerable help in the prevention of accidents.

When a birdcage scaffolding supports a soffit formwork for a bridge deck or similar structure, adequate diagonal bracing is necessary to improve the stability of the scaffolding. Vibration, impact, and inse-

Plate 11.2   Poker vibrators for concrete. Courtesy Wacker Canada Ltd., Mississauga, Ont., Canada.

cure ground support may all contribute to scaffolding failure. The vibration may dislodge the timbers, adjustable forkheads, or scaffolding out of line, thus inducing a failure. In a similar way the impact from the accidental blow of site machinery may displace the formwork supports. Ground conditions may be improved for formwork supports by compaction, stabilization by lean mix or a layer of imported fill, or the laying of large timbers, such as railway sleepers, so that the loads from the concrete and formwork are evenly distributed over the ground. The deleterious effect of a thaw on frozen ground should be closely scrutinized when conditions are appropriate. It is shown in Section 11.4 that a drop in temperature can substantially increase the concrete pressure. In certain circumstances on site it therefore might be necessary to reduce the rate of pour. The importance of the rate of pour on the concrete pressure should be understood by site personnel.

Plate 11.3 Formwork vibrators. Courtesy Wacker Canada Ltd., Mississauga, Ont., Canada.

It is possible that despite adequate formwork design a small error on site can cause local overstressing, making the formwork dangerous. Adequate practical and technical supervision and inspection should prevent this.

Premature release of formwork and the mistaken sizes of materials and equipment used in formwork construction have been fairly common reasons of formwork failure in the past. The planning and provision of sturdy and safe access for all site personnel can greatly improve the prospects of safety. Toe boards, safety rails, the use of helmets, and the barricading of areas where formwork is likely to fall when being stripped are items that the contractor can plan and provide for, and the engineer has an obligation to provide an adequate system of inspection of these.

Plate 11.4 Formwork vibrators. Courtesy Wacker Canada Ltd., Mississauga, Ont., Canada.

Plate 11.5 A concrete agitator truck. Courtesy Winget Ltd., Rochester, Kent, England.

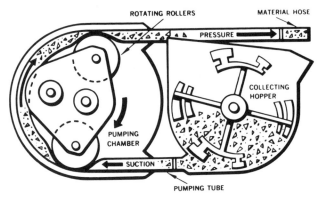

Plate 11.6   Concrete pumping equipment. Courtesy Canadian Portland Cement Association, St. John's, Canada.

Plate 11.7   A slump test. Courtesy Canadian Portland Cement Association, St. John's, Canada.

11.11   CONCRETING IN COLD WEATHER

The variation in concrete pressure with change of temperature is shown in Section 11.4.

Adequate provision must be made to prevent rapid temperature changes in concrete that has recently been cast. During the first few

Plate 11.8 Making a test cylinder. Courtesy Canadian Portland Cement Association, St. John's, Canada.

Plate 11.9 Testing a test cylinder. Canadian Portland Cement Association, St. John's, Canada.

days the heat of hydration of the hardening cement is developed. Usually little heat from outside sources is required if the heat generated within the concrete is adequately conserved. The allowable rate of change of temperature depends on the thickness and size of the concrete sections.

Several types of insulation are available. Polyurethane or polystyrene foam boards may be attached to the outside of the plywood and slotted between the bearers. There are also various types of blanket insulation. Electric heating blankets are available to ensure that the temperature does not drop too quickly. Salamanders or other types of oil or electric direct heat appliances are useful on structures where tarpaulin enclosures are possible.

Many ready mixed concrete plants have heating facilities that enable concrete production to continue throughout the winter months except in areas where very severe winter weather is common.

Concrete can be placed safely during the winter months if precautions are taken. Special protection must be provided when low temperatures occur, particularly during the placing and early curing period. The permissible minimum concrete temperature is usually considered to be 5°C, and protection from frost should be maintained until the concrete attains a compressive strength of about 7 N/mm^2. Concrete that is allowed to freeze soon after placing gains very little in strength, and permanent damage is certain to occur.

11.12 CONCRETE TESTS

Tests are usually carried out on a randomly selected batch of concrete to check that the concrete maintains a consistency during a concrete pour. At fairly frequent intervals control tests are made on the amount of slump. The slump test for consistency of concrete should be made in accordance with the specification. The slump cone must be filled in three layers of approximately equal volume, after which the cone is removed and the slump measured.

To check the variability of strength in batches, concrete cylinders are made. The standard test specimen for the compressive strength of concrete with a maximum aggregate size of 40 mm or less is a cylinder 150 mm in diameter by 300 mm high. Other types of test specimens include a 100×200 mm cylinder, and 100 and 150 mm cubes. Cubes exhibit higher compressive strengths than cylinders. Tests are made after 28 days, although accelerated curing tests can be used to expedite the results from quality control. Test cylinders and cubes are filled and rodded in three approximately equal volumes.

Occasionally beams are made for flexure tests with 150×150 mm cross section and about 450 mm long. There are many more tests available to the engineer so that a good quality control can be maintained, such as tests for tensile strength, organic impurities, grading of aggregate, temperature measurement, air content, density, and cement content.

11.13   WORKED EXAMPLES

Note that Examples 11.1 and 11.2 have been inserted at specific points within the text, not at the end of the chapter, to facilitate the learning process and maintain continuity.

Example 11.3

A concrete wall is to be 4 m high and 150 mm thick. It is anticipated that the rate of placing will be 2 m/hr, the slump will be 50 mm, and the concrete temperature will be 15°C. Determine the design pressure using equations 11.1 to 11.3. No allowance need be made for impact of the concrete.

Answer 11.3

By height:

$$P_{max} = \Delta H$$
$$= 23.6 \times 4$$
$$= 94.4 \text{ kN/m}^2$$

Arching effect:

$$P_{max} = 3R + \frac{d}{10} + 15$$

$$= 3 \times 2 + \frac{150}{10} + 15$$

$$= 36 \text{ kN/m}^2$$

Stiffening effect:

$$P_{max} = \Delta RK + 5$$
$$= 23.6 \times 2 \times 1.1 + 5$$
$$= 56.9 \text{ kN/m}^2$$

Therefore the design pressure is
$$P_{max} = 36 \text{ kN/m}^2$$

Example 11.4

Repeat Example 11.3 using Figure 11.5.

Answer 11.4

By height:
$$P_{max} = 92 \text{ kN/m}^2$$

Arching effect:
$$P_{max} = 35 \text{ kN/m}^2$$

Stiffening effect:
$$P_{max} = 54 \text{ kN/m}^2$$

Therefore the design pressure is
$$P_{max} = 35 \text{ kN/m}^2$$

Example 11.5

Repeat Example 11.3 using Tables 11.3 to 11.5.

Answer 11.5

By height:
$$P_{max} = 100 \text{ kN/m}^2$$

Arching effect:
$$P_{max} = 35 \text{ kN/m}^2$$

Stiffening effect:
$$P_{max} = 55 \text{ kN/m}^2$$

Therefore the design pressure is
$$P_{max} = 35 \text{ kN/m}^2$$

Example 11.6

Repeat Example 11.3 using equation 11.4 or 11.5.

Answer 11.6

	Imperial units	SI units
Height	13.12 ft	4.0 m
Thickness	6 in.	150 mm
Rate	6.56 ft/hr	2 m/hr
Slump	2 in.	50 mm
Concrete temperature	59°F	15°C

As the rate of pour does not exceed 7 ft/hr, equation 11.4 should be used:

$$P_i = 150 + \frac{9000\,R_i}{T_i}$$

$$= 150 + \frac{9000 \times 6.56}{59}$$

$$= 1150.7 \text{ psf}$$
$$= 55 \text{ kN/m}^2$$

In this calculation the slenderness of the wall is not accounted for, and therefore the reduction of pressure arising from the arching effect is not considered.

Example 11.7

A column is 3.5 m high and 300 × 300 mm in cross section. If the rate of placing is 15 m/hr and the concrete has a temperature of 20°C and a slump of 100 mm, determine the maximum pressure on the formwork using Tables 11.3 to 11.5. No allowance need be made for any impact on the formwork.

Answer 11.7

By height:

$$P_{max} = 87 \text{ kN/m}^2$$

Arching effect:

$$P_{max} = 90 \text{ kN/m}^2$$

Stiffening effect:

$$P_{max} = 150 \text{ kN/m}^2$$

Therefore the design pressure is

$$P_{max} = 87 \text{ kN/m}^2$$

Example 11.8

Determine the spacing of the studs and the spacing of the tie bolts in a formwork panel if the maximum concrete pressure is 50 kN/m². Assume that the plywood is 19 mm thick and the working stresses are as follows:

$$\text{Bending} = 10 \text{ N/mm}^2$$
$$\text{Shear} = 1.1 \text{ N/mm}^2$$
$$E = 6.9 \text{ kN/mm}^2$$

Answer 11.8

Using equation 11.8 for the bending check yields

$$l = \sqrt{\frac{fbh^2}{0.6w}}$$

$$= \left(\frac{10 \times 10^6}{10^3} \times 1 \times \frac{19^2}{10^6} \times \frac{1}{0.6 \times 50} \right)^{1/2} \quad \text{(kN-m units)}$$

$$= 0.3468 \text{ m}$$
$$= 347 \text{ mm}$$

Using equation 11.11 for the shear check yields

$$l = \frac{1.11 \, bh s_{max}}{w}$$

$$= \frac{1.11 \times 1.0}{50} \times \frac{19}{10^3} \times \frac{1.1 \times 10^6}{10^3} \quad \text{(kN-m units)}$$

$$= 0.464 \text{ m}$$
$$= 464 \text{ mm}$$

Assume a stud spacing of 350 mm center to center, that is, a clear span of, say, $350 - 50 = 300$.

Using equation 11.12 for a deflection check gives

$$\delta_{max} = \frac{kwl^4}{EI}$$

$$K = 0.0069 \text{ for continuous beams}$$

Therefore

$$\delta_{max} = 0.0069 \times 50 \times 0.35^4 \times \frac{1}{6.9 \times 10^6} \times \frac{12 \times 10^9}{1 \times 10^3}$$

$$= 0.00131 \text{ m}$$

$$= 1.31 \text{ mm, using the center to center span}$$

or

$$\delta_{max} = 0.71 \text{ mm, using the clear span}$$

The deflection is therefore less than 2 mm but is not less than 0.003 times the center to center span; that is, $0.003 \times 350 = 1.05$ mm, which is less than 1.31 mm. However, it should be noted that the deflection over the clear span is only 0.71 mm, which is well within the required limits. This large difference in deflection due to which span is considered frequently occurs in formwork design where the spans are very short, and the deflection is proportional to the span to the power 4. Whether the span should be reduced is therefore a difficult engineering decision. First, if the concrete is to be covered by an architectural finish or is not to be frequently seen by the public, the spacing of the studs can remain at 350 mm center to center. Second, the end spans only could be reduced to reduce the maximum moments and deflections. Third, if the spans were reduced to 325 mm, the center to center span deflection would be reduced to 0.97 mm, which would equal 0.003 times the span value.

If a bolt diameter of 17.5 mm is used, see Table 11.6, the maximum safe working load is 57 kN. Therefore the spacing of the tie bolts is $57/50 = 1.14$ m²; that is, one tie bolt is required for every 1.14 m² of formwork.

CHAPTER 12

TUNNEL ENGINEERING

12.1 INTRODUCTION

Tunnels are used for the underground movement of road and rail transportation and pedestrians. They are also used in mining and hydro schemes and can be used for drainage and other services.

The different types of tunnels are shown in Figure 12.1. Tunnels in water may be constructed by using cofferdam or immersed tube techniques. When tunnels are required at a subsurface level, the cut and cover method is used, provided that the proposed line of the tunnel is reasonably accessible. Usually in the cut and cover method the normal type of excavating equipment can be used from ground level or on benches below.

For deeper tunnels in relatively soft soils tunneling machines or shields may be used. A recent advance in tunnel engineering is the use of tunneling machines in fairly hard rock instead of the use of drill carriages and explosives. The following sections describe these different methods and types in more detail.

Because of the importance of the geological conditions within the area of the proposed tunnel, a detailed accurate survey and site investigation is necessary to assess the feasibility, type of tunnel, and method of construction. This is not always possible for tunnels in subaqueous or mountainous regions, where geophysical methods are used.

The porosity of the soil and frequency of faults and joints should be known before tunneling commences. Small diameter pilot tunnels are sometimes used to obtain precise geological information along the line of the tunnel. If the ingress of water is likely to be a problem, the following techniques are available to the tunnel engineer:

1. The use of compressed air
2. Dewatering by well points

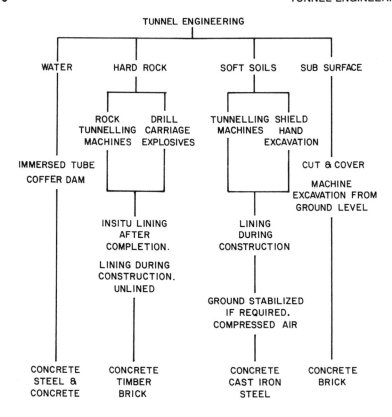

Figure 12.1   Types of tunnels.

3. Stabilizing the ground by cement or chemical injection
4. The use of bentonite
5. Freezing the ground

12.2 GEOLOGICAL CONDITIONS

The geological conditions along the proposed line of a tunnel are extremely important. Tunnels are driven through all types of rocks within the three rock groups: igneous, metamorphic, and sedimentary.

Faults should be avoided, if at all possible, when driving tunnels through the geologically older massive types of rock because shattered broken rock, called fault breccia, is often associated with fault planes. Apart from the necessity of extra supports or stabilization, water is very likely to seep through the brecciated areas in unknown quantities.

The amount and size of joints present in the rock determine the amount of overbreak (see Section 12.4).

Although much greater production rates can be achieved in geologically younger soils, because of the use of tunneling machines, hazards are generally more frequent. The presence of water in sands, silts, clays, and alluvium can create serious problems for the tunneling engineers. Because of the growth of many cities around river estuaries these geological conditions are quite prevalent in many parts of the world where a new tunnel is proposed. About a century ago disasters were all too frequent in the early stages of tunneling development in these types of geological strata.

12.3 TECHNICAL ASPECTS

Today the art of designing tunnels is still largely based on previous experience and certain basic assumptions in the theoretical aspects. For many years tunnel engineers have attempted to ascertain the loading conditions on tunnel linings. Because of the lack of uniformity of the earth pressures arising from nonuniform soil properties along the entire length of the tunnel lining, and because the methods of measuring data on existing tunnels have not been well established and have not been universally well documented, the design techniques and criteria available to the engineer are still very limited. However, it is difficult to imagine that a rigorous theory for the assessment of earth pressures could be used accurately for general use because of the many variables. For instance, in underground railway systems the earth pressures themselves are often time dependent because of the heating up and drying out of the surrounding soil over a period of years.

The principle of the design of flexible linings in suitable soft ground is that any inequality between the vertical and the horizontal loading on the lining causes the lining to deform until sufficient passive ground resistance is mobilized to give a roughly uniform radial pressure all around the tunnel lining. In practice, linings, whether segmentally bolted or not, generally deform and behave as flexible linings. The deformation may be calculated as described by Morgan[26] from a known loading condition. This calculation confirms that deformation is resisted almost entirely by the mobilization of the passive pressure and that the resistance provided by the bending stiffness of the segments is small.

Generally in hard rock in situ concrete is used for the tunnel lining. This creates a rigid lining. The effects of faults and weak planes in the rock create great variation in the loading conditions along the line of a tunnel. Again, the right type of instrumentation and documentation of existing tunnels has not been available to produce relevant data on which a rigorous design may be based.

12.4 DRILLING AND BLASTING

In recent years tunneling machines have been developed which are capable of boring in fairly hard rock. This method has tended to replace the older methods of drilling and blasting. However, there are still many cases where it is necessary to use blasting techniques to excavate rock for tunnels. It has been suggested[27] that tunneling machines using drag picks could not bore into the harder igneous rocks but could be used on many sedimentary rocks up to a hardness of about 125 N/mm^2. Tests carried out in Germany[28] showed that a practical limit for machines equipped with tungsten carbide cutters gives a limiting value for F, the rock abrasion coefficient, of 500 N/m, and for those with tricone roller bits, 2.7 kN/m, although the fragmentation action of the types of cutter and cutter heads affects the limiting values. However, it is obvious that there is some limitation on this type of machine because of the hardness or crushing strength of different rocks. As time progresses, owing to the further development of the rock tunneling machines, these maximum limiting values will probably increase.

Even in tunnels where the rocks are not too hard for the utilization of tunneling machines the length of the tunnel, if short, may make the use of blasting techniques a more economical method. Obviously the initial cost of the installation of a tunneling machine is prohibitive over short tunnel lengths. Vertical shafts, whatever the rock strength, require a drilling and blasting method. Drilling and blasting are more labor intensive and, coupled with the reduction of overbreak and saving in concrete with tunneling machines, the continued development and expansion of the use of machines are likely.

The most common method today of drilling and blasting is by drilling holes in the full face of the tunnel cross section to a set pattern, with the use of delay detonators during the blasting stage. The holes can be drilled by a drill carriage or jumbo. Modern jumbos, consisting of any combination of hydrabooms or parallel booms mounted on an undercarriage, are either self-propelled, truck mounted, track mounted, or rail mounted units with up to about seven booms to give maximum face coverage and fast drilling time.

The amount of explosive, which should be ascertained by trial, varies, as does the size of the holes, according to the nature of the rock, local conditions, and the type of explosive. The pattern and spacing of the holes also vary depending on the required degree of fragmentation of the rock. The optimum pattern size is usually determined by experience or experimentation with each change in the rock formation. As in all blasting work, control and economy are brought about by attempting to arrange for the rock to be blasted toward a free face. It can be seen in Figure 12.2 that if the cone shaped volume of rock is first removed, subsequent volumes around one or more concentric rings of easer holes may be removed far more easily by blasting. This operation may be performed by the use of millisecond delay detonators.

After all the holes have been drilled and loaded with explosive, the lines of cord are laid out and millisecond delay detonators are installed. Delay detonators interrupt the normal speed of detonation of the lines by a specific time lag. Delay times may vary from 10 to 70 msec. The position of the delay connectors determines which holes are detonated at one particular instant of time. The delays enable each ring to be blasted essentially toward a free face, controlling and easing fragmentation of the rock, after the first ring of cut holes has created a conical space. Delays also prevent the blasting operation occurring at one instant, thereby reducing air blast and ground vibrations.

Very tough rock, such as granite or gneiss, can be advantageous insofar as little bolting or grouting is necessary. If the tunnel[29] is not being lined, a smooth edge may be achieved by using lighter explosives, for presplitting, for those charges closest to the edge. In this

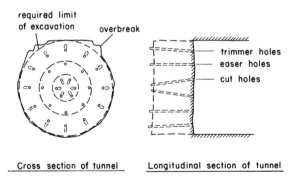

required limit of excavation   overbreak

trimmer holes
easer holes
cut holes

Cross section of tunnel   Longitudinal section of tunnel

Figure 12.2 A hole pattern for blasting in tunnels.

method, however, the outer edge is detonated first to separate the rock to be removed from the wall. Phased blasting over a millisecond period is used to give blast waves an easy lateral path, as well as the natural one back down the tunnel. If water is to flow in the tunnel and the tunnel is not being lined, the tunnel still requires a smooth wall to create a smooth water flow.

Overbreak is the amount of rock that is removed beyond the required excavation limits. Refer to Figure 12.2. The amount of overbreak depends on the blasting operation, the type of rock, and the angle of dip in the rock in relation to the line of the tunnel. Payment for excavation is only made for excavation up to the required limits. Overbreak, although inevitable when blasting techniques are used, should be kept to an absolute minimum by continuously reviewing blasting operations. Enormous extra expense can be incurred by the contractor by not giving special attention to the amount of overbreak. For instance, a 250 mm average overbreak on a 5.000 m diameter tunnel requires about a 20% increase in volume of rock disposal and about a 100% increase in concrete requirements for the lining.

The disposal of rock, frequently called the mucking operation, is usually done by a belt conveyor system or small rail trucks.

To maintain clean fresh air in the tunnels for all personnel, extensive ventilation systems need to be installed. These systems are usually two way control systems which also remove dust and noxious fumes. An air replacement rate of 7 to 15 m^3/min per person is usually required to maintain adequate ventilation.

12.5 TEMPORARY AND PERMANENT TUNNEL FORMWORK

An in situ concrete lining is one method of lining the tunnel after the rock has been removed. Proprietary traveling formwork such as the one shown in Figure 12.3 is used to place the in situ concrete. The steel face, stiffened by webs, is mounted on a steel frame which in turn is usually mounted on rail tracks. Adjustable jacks between the top of the formwork and the frame enable the formwork to be raised exactly to the correct position for the in situ concrete. The side jacks enable the sides of the formwork, which pivot at the top, to be placed exactly in position.

Frames at each end of the formwork are connected by longitudinal steel members. Holes are left in the steel face to enable vibrators to be inserted into the freshly pumped concrete. As the concrete is being placed, the flow of concrete and reinforcement can also be inspected

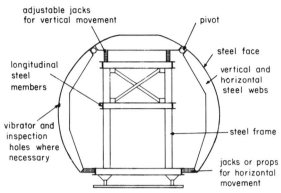

Figure 12.3   Traveling formwork for tunnel linings.

through the holes. When one section of the tunnel lining is complete, the sides are moved inward and the top part of the formwork is lowered so that the formwork can be moved quite easily and quickly to the next section.

Another method of construction, where the rock has a limited period of stability, is to underpin the upper part of the tunnel excavation with a concrete arch within the shortest possible time after excavation. Single concrete rings are installed individually at the same rate as excavation proceeds and immediately following the tunnel advance.

One particular application of this type of tunnel construction is known as the Bernold[30] system. Specially shaped, perforated ribbed steel sheets are used for the dual purpose of a permanent formwork and a reinforcement cage. The sheets are joined together and are temporarily supported on assembly arches. Because of their special shape the sheets are rigid enough to resist the concrete pressure and are also sufficiently impervious to prevent a stiff plastic concrete from running through the perforations during vibration of the concrete. The concrete may be pumped in or sprayed through the sheets. Figure 12.4 shows a particular use of the Bernold system.

The system fulfills two different functions, namely, that of the steel supporting the frame and that of the permanent rock lining. The concrete placed between the sheets and rock face protects the face from the effects of air and moisture and supports the rock, preventing further fragmentation. Depending on the stability of the rock the unsupported length of tunnel usually varies between 2.0 and 3.0 m. The zone immediately behind the advance in the tunnel is supported in about 5 hr, depending on the drilling and blasting program. This program also determines the length of the concreting operation.

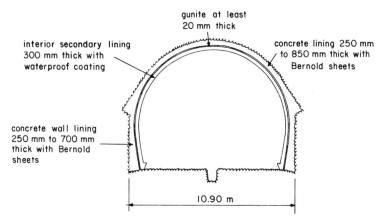

Figure 12.4   The Bernold system of tunnel construction.

Figure 12.4 shows how the Bernold system is incorporated into the permanent structure.

12.6 SPRAY-ON CONCRETE

When the overburden is about 1000 m or more, tremendous rock bursts can occur in tunnels because of the adjacent rock expanding into the tunnel. This local stress relief causes pieces of rock around the tunnel perimeter to burst into the tunnel opening. It was found[31,32] that a thin concrete lining, sprayed on the rock soon after blasting, helps prevent rock bursts.

The concrete lining can be sprayed onto the recently exposed surface at a certain distance behind the blasting face to prevent rapid rock decomposition. The method is usually referred to as the Shotcrete method, after its trade name.

It is possible to spray the concrete immediately after blasting and, if necessary, the Shotcrete can be strengthened with steel arches and mesh reinforcement. The development and improvement of the machines and equipment for spraying concrete have been rapid. Coarse aggregate Shotcreting is commonly used today. Rock bolts of various types can be installed as additional support to improve the effectiveness of the Shotcrete.

This method requires very skilled and experienced engineers and operators, as the success of the lining operation depends on the installation time and the properties of the rock. It is probable that the use of the spray-on method will increase with the development of the coarse aggregate technique and the use of accelerators in the wet mix Shotcrete.

12.7 SUPPORT SYSTEMS AND SHIELDS

For tunneling in rock there are several methods of temporarily supporting the rock face until the permanent lining is installed.

Steel arches can be laid on timber blocks and brought into contact with the rock by timber packing pieces. Once the timber has been compressed, the steel arch props up the rock unless the rock is incapable of supporting the load, in which case the load is transferred to the steel arch, which deforms and redistributes the load to the less highly stressed areas in the rock. Generally the highest stress is in the roof or soffit and is redistributed to the walls.

Another method of supporting the rock face is by the use of rock bolts. The rock bolts are anchored into the rock and loaded in tension. By transferring the load, the rock is placed in compression in a similar way to prestressed concrete principles. A zone of compression is created by a pattern of rock bolts and the rock stresses assume a state of equilibrium. Anchorage length, diameter, and spacing of the bolts are important factors in determining the effectiveness of rock bolts.

Because of the unpredictable nature of the rock encountered frequently in tunnels, rock bolts can be useful in local confined areas of the tunnel which are not as stable as most of the tunnel length. If the roof falls are severe, an in situ concrete roof can be installed to support the roof in conjunction with the rock bolts. The concrete can be placed by using the canopy of the tunneling machine as formwork if it is present.

The cast in situ concrete method all along the tunnel length is an extremely common method of lining tunnels permanently. When a near state of equilibrium has been achieved by the use of temporary supports, the concrete is placed and, after it has achieved sufficient strength, supports the tunnel loading. The Shotcrete method described in the preceding section can simplify this operation.

In soft ground tunneling work a shield is used. A shield supports the sides of a tunnel before the ring of tunnel lining has been placed. It is used for driving tunnels in soft ground with diameters from 2.0 to 12.0 m. Mechanical cutters may be incorporated with the shield to form what is called a tunneling machine. These machines are described in Section 12.9. When mechanical cutting means are not used, the excavation is done by miners using mechanical hand tools. A steel cutting edge, often with a hood at the top to ensure greater protection for the miners, is mounted at the front of the shield. Hydraulic jacks, mounted inside the circumference of the skin of the shield, push against the last assembled ring of lining and move the shield forward a distance equal to the width of the lining, which is usually about 500 to 800 mm. After further excavation the rams are retracted and the next ring of lining is placed. The skin of the shield is often just greater than the outer diameter of the lining, and the skin continues to support the ground while the new ring of lining is placed. Later cement grout is injected through specially prepared holes in the lining to fill the space between the ground and the lining. However, if the ground can support itself temporarily, such as a stiff clay, the lining can be placed directly against the ground and a shorter shield can be used.

The shield may consist of a fully welded steel plate or segmental units which are erected in underground chambers. It should have a very smooth outside surface to facilitate driving. The main thrust rams must be capable of overcoming the skin friction of the shield surface and the resistance of the cutting edge thrust.

Some shields may contain several cells to enable excavation to proceed in several areas. Another type of shield contains a diaphragm with doors at apertures, so that miners in the shield can be isolated from the face of the tunnel where the ground conditions are poor. For large tunnel diameters the shields require internal framework to support the skin.

12.8 LININGS

There are three common types of tunnel linings: in situ concrete, Shotcrete, and the prefabricated or precast segmental type of lining. The in situ concrete and the Shotcrete type of linings are described in Sections 12.5 and 12.6. The segmental type of lining is now usually precast concrete, but cast iron segments are still used in tunnel lengths where, for instance, nonstandard diameters are used. About 40 years ago the precast concrete segmental lining began to replace cast iron linings, mainly because of cost. Subsequently the precast concrete lining has been further developed. A new type of precast concrete lining was developed where the segmental units are placed directly against the ground, when suitable, and expanded into place by wedged segments. This method leaves the excavated ground unsupported for the least possible time and reintroduces into the ground some of the stress re-

leased on excavation. There are now several varieties of concrete linings, some of which are jacked onto the tunnel circumference by rams
incorporated in the tunneling machine. In noncohesive water bearing
soil cast iron linings are still the safest, although in general expanded
into place linings, built behind a shield or tunneling machine cutting a
circular profile, are the safest, fastest, and most economical method of
tunneling today.

If the ground is not temporarily self-supporting, bolted linings, built
within the tail of the shield, must be used.

Cast iron linings may be bolted or expanded into position. The bolted
type provides additional stiffness and longitudinal continuity. A bolted
type of lining is shown in Figure 12.5. These segments are usually
cement grouted and caulked. A flexible jointed cast iron lining is
shown in Figure 12.6.

One type of precast concrete lining is shown in Figure 12.7, although
several types are in use today. Flexure is possible because the lining
units are able to pivot at the joints. A small reinforcement cage may be
added to the female joint to prevent spalling. The jacking of the units
together is usually carried out at about a third of the height of the
tunnel. Sometimes the rings are retained in the fully jacked position by
the insertion of tapered jacking pieces. The jacks are then withdrawn
and their pockets filled with a precast block and expanding mortar.

After each tunneling project is completed, a minor modification may
be made to a precast concrete lining for a subsequent project. It is for
this reason that Figures 12.5 to 12.7 cannot purport to show the latest

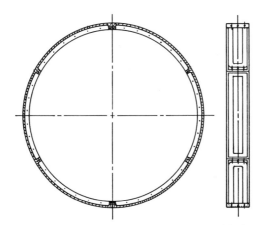

Figure 12.5 Bolted type of cost iron tunnel lining.

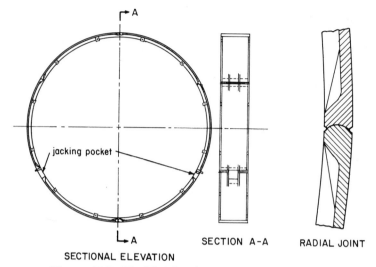

SECTIONAL ELEVATION   SECTION A-A   RADIAL JOINT

Figure 12.6   Flexible jointed cast iron tunnel lining.

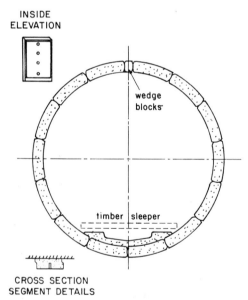

CROSS SECTION
SEGMENT DETAILS

Figure 12.7   Flexible precast concrete lining.

type of lining and should therefore be regarded as showing only the general details of the different types of tunnel linings.

Neither cast iron nor precast concrete linings are suitable in situations where high tensile stresses are likely to be induced in the lining. In this case steel linings are an alternative, but these are very expensive even when the cost of corrosion is neglected. Investigations have been made to consider the feasibility of using ductile iron segments. With recent metallurgical advances it is possible that in the future ductile iron may become a viable alternative choice.

12.9 MACHINE TUNNELING IN SOFT GROUND

The most suitable type of ground for mechanized tunneling is firm, dry clay or soft, unfissured rock such as chalk, sandstone, or limestone. Tunneling machines are made in parts so that they may be transported in suitable sizes and erected in the tunnel.

The principal factors affecting the speed of construction of a tunnel in clay with a 4.0 m diameter, for example, are the actual speed of excavating the face, the speed with which the lining segments can be erected and fixed, and the rate of grouting behind the completed lining section.

Since modern tunneling machines in soft ground are equipped to excavate the face and erect the lining segments, the bolting and grouting of the segments are now frequently not necessary, which accounts for a 25 to 30% saving of the total cycle time.

The modern tunneling machines contain a shield, described in Section 12.7, and a cutting head attached at the leading end. The cutting head consists of steel digging teeth mounted on a cutter arm that rotates in front of the machine. Hydraulic jacks, which are mounted inside the perimeter of the skin of the shield, push against the most recently erected and positioned ring of lining and move the machine forward a distance equal to the width of the lining. The lining must therefore be designed to take this longitudinal force, with respect to the tunnel axis, induced by the jacks during construction. The jacks or rams are withdrawn, and the ring of lining is jacked into position by another system of jacks located at the back of the machine.

Two types of tunneling machines are shown in Figure 12.8. One type contains a rotating drum which incorporates the digging teeth and the other type contains cutter arms with digging teeth attached to a central shaft.

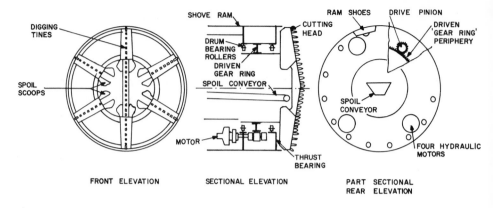

(a) A Drum Type Machine

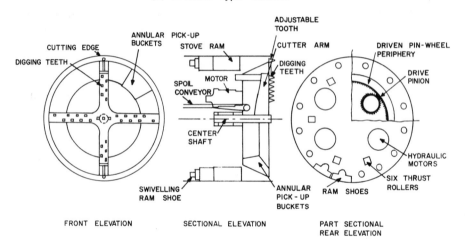

(b) A Center Shaft Type Machine

Figure 12.8 Tunneling machines for soft ground.

For small diameter tunnels the traditional center shaft and drive machinery creates obstruction to spoil extraction. The alternative is to use compact hydraulic drive units, sufficiently small to be mounted around the circumference of the machine, to drive the periphery of the cutter. The central shaft is then replaced by peripheral bearings. Comparisons have been made[33] which indicate that the shaft type of machine appears superior beyond a certain tunnel diameter and that there is a limiting diameter below which the drum digger provides the only practical answer.

The design of machinery and equipment following the digging unit, for handling and disposal of the spoil and for segment lining erection and positioning, depends on the rate of progress of the tunneling machine, which is in turn dependent on the cutting speed. The cutting speed is determined by the type of soil, rate of spoil removal, the type and configuration of the teeth, the maximum longitudinal force that the lining segments can withstand, and the rotational power available for the cutting head. Obviously most of these factors are interrelated. The removal of spoil is dependent on the rate of loading at the face, the speed of transportation, and the removal at the surface.

In the United Kingdom McAlpine has developed a machine that is designed to be capable of working in either soft or hard ground. When the machine is operating in hard material, steering skis on the shield are extended and a fulcrum plate around which the machine pivots is attached at the front. The ski panels are not gripper pads. Thrust is obtained through rams around the shield, positioned in a similar way to the soft ground machine, which push against a reaction ring clamped against the sides of the tunnel. The development of this type of machine, which is capable of driving in soft and hard ground, is continuing.

In spite of the high rates of production that can be obtained from a tunneling machine it only becomes economical if the planned tunnel lengths exceed 1.5 km. For shorter lengths a shield and the use of hand tools are the most economical method. Although the running costs of a machine are cheaper than manual methods, the machine installation and site costs are greater at the beginning of the project. The break-even point depends on the type of rock and many other variables, but is usually considered to be at about 1.5 to 2.0 km.

12.10 MACHINE TUNNELING IN HARD ROCK

Machines are not capable of tunneling in the harder igneous rocks but are capable of tunneling in many hard sedimentary rock formations up to a hardness of about 125 N/mm^2. The rate of progress with this type of machine is mainly governed by the hardness of the rock. In the traditional method of drilling and blasting the rate of progress is often little different between hard and softer rocks, so rock tunneling machines are at their most advantageous in softer rock, particularly since considerably fewer temporary supports are required when machines are used.

Machine tunneling in hard rock has several advantages over drilling and blasting. There is less overbreak and therefore the lining cost is very much reduced. The number of skilled men is reduced. Safety is increased. The excavation and removal of spoil are almost continuous and a faster rate of drive is accomplished. Some modern machines use a small rotating cutting head of about 1.0 m diameter which traverses along an arm that rotates about a horizontal shaft. It is therefore possible to cut an arch profile. The cutting drum is initially thrust into the middle of the face of the rock and then traversed radially outward while rotating on the arm. All cutting faces are therefore radial after the initial thrust, and the break out of the rock is in tension to a free face. The fact that the machine does not cut out the full face, because the diameter of the cutter is much less than the tunnel diameter, means that the axial force on the machine is very considerably reduced. Progressive stages in the cutting cycle are shown in Figure 12.9.

To drill the whole face at a time, with disk cutters mounted across the full diameter, requires a reaction of approximately 3500 kN to resist the thrust whereas with the small cutter diameter, as mounted on the Greenside-McAlpine machine,[34] the thrust required is only about 120 kN to enter the cutter head into the rock. Even with a four cutter machine on a tunnel of about 7 m diameter, the total thrust requirement is only about 500 kN.

The disadvantage of a high thrust machine is that the reaction must act against temporary or permanent lining and possibly include a special expanding thrust member, although a machine excavating over a full face in the softer types of hard rock can obviously achieve a higher production rate. High pressure water jets applied in conjunction with roller cutters on a full face tunneling machine can either reduce the thrust required or increase cutter speed for the same cutter costs.

The rock cutter bits are very expensive and it is important to have

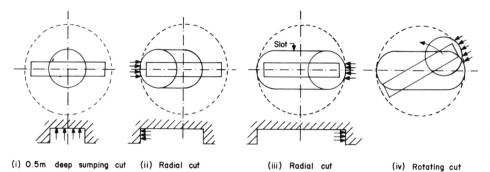

(i) 0.5m deep sumping cut   (ii) Radial cut   (iii) Radial cut   (iv) Rotating cut

Figure 12.9   Rock tunneling machine cutting cycle.

the correct shape of angle of attack and relief to the cutting edges. The picks are mounted on steel shanks so that they can be replaced quickly and tungsten carbide tips can be inserted into the drag pick type cutter.

Two or four rotating cutter heads with radially projecting teeth can be mounted on the main drum so that their axes of rotation are slightly out of line with the tunnel axis and the drum axis. The rate of progress is slower than the full face machines cutting the whole face. Modern rock tunneling machines that cut the full face using inclined roller bits are being developed, and these incorporate gripper pads instead of expanding thrust members.

In the Greenside-McAlpine rock tunneling machine the machine base controls the separate operations of horizontal and vertical movement during advance while negotiating curves or accomplishing gradient changes. These operations are achieved by using hydraulic jacks to lift the machine while rams push it forward. Figure 12.10 diagrammatically demonstrates the walking base method. The walking floats slide under the body of the machine, the body of the machine is lifted by the jacks, the ram is retracted, and the body of the machine moves forward. The body is then lowered down by the jacks into the cutting position.

Steering alignment is achieved by a laser beam. The canopy provides forward area roof support and, when positioned in place against the tunnel walls, stabilizes and protects the machine while it is working. It also houses the dust extraction system (700 m³/hr), which is connected

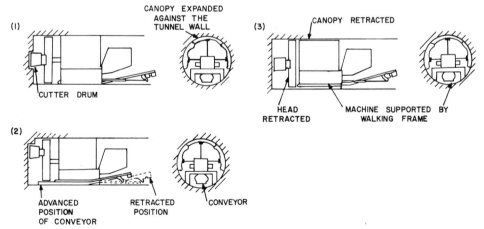

Figure 12.10 Advance movement of rock tunneling machines.

to the surface by ducts inside the tunnel. The canopy can be raised and lowered hydraulically by top and side rams.

Excavated material is removed by a conveyor which can be hydraulically advanced or retracted. Two hydraulically operated loader blades guide the spoil material onto the conveyor, which operates at a speed of about 60 m/min. The spoil from the cutters is a reasonable size for handling and is not ground to dust. Blades scrape the spoil material from the tunnel base to prevent a tendency of the machine to climb on excavated material as it advances.

The machine is designed to excavate and remove at the rate of 20,000 kg/hr in medium to hard rocks of unconfined compression strengths of up to 150 N/mm^2. The peripheral speed of the pick cannot be increased beyond a certain speed to increase production because of the heat buildup during the cutting portion of the cycle. The cutting procedure is so arranged that the pick is only in intermittent contact with the rock face. It is therefore better to use the power of the machine for greater pick penetration at a slower speed rather than lesser penetration at a higher speed.

The cutter cost is an important item in the estimate of tunneling machine costs. One parameter, the rock strength, cannot alone determine the rate of progress and cutter cost. The quartz content in the rock and grain size also affect the cutter wear. It has been mentioned that if F, the rock abrasion coefficient, is greater than 500 N/m for tungsten carbide cutters or 2.7 kN/m for tricone roller bits, a machine tends to become impractical. The rock abrasion factor is

$$F = \text{quartz content} \times \text{grain size} \times \text{tensile strength}$$

A more economical production rate is possible in the future by the use of a more powerful machine and a better knowledge in pick shape, metallurgical composition, and reuse after grinding. Recently full face machines have excavated in sedimentary rocks varying in hardness from 20 to 200 N/mm^2 and even in thin layers of igneous rocks up to a hardness of 450 N/mm^2.

12.11   THE BENTONITE TUNNELING MACHINE

Some cities are built on rock or clay and it is common to find a well developed systems of tunnels and subways in these localities, but many others sited on alluvial deposits have had such construction held back by the high cost of tunneling in these soils.

There are a number of methods of tunneling in sands and gravels. If, however, a tunnel is being constructed below the water table in poor ground, it is usually necessary to use additional means to prevent the tunnel face from collapsing. Air under pressure can be maintained within the tunnel with an air lock at the entrance. Alternatively or additionally the ground through which the tunnel is to be driven may be stabilized by cement or chemical grouting. In certain circumstances the ground may be stabilized by freezing. These methods are described more fully in Section 12.12. They are generally fairly slow and costly, and the method may not only be a health hazard but also dangerous as in the case of compressed air.

The Nuttall-Priestley Bentonite tunneling machine[35,36] presents an economic and safer alternative. The machine is capable of driving tunnels through granular soils above or below the water table. A cutter head revolves in a sealed chamber that contains a thixotropic slurry, called Bentonite, under pressure which supports the tunnel face. The machine is shown diagrammatically in Figure 12.11.

The thrust rams, located behind the pressure bulkhead, are capable of providing the thrust necessary for digging and also overcoming the pressure of the slurry in the front chamber. The tunnel lining is erected in the usual manner, using hydraulically operated gear within the tail of the machine. Incorporated in the erection gear are hydraulic struts for holding the machine stationary against the pressurized slurry. After the machine has advanced a full stroke and the lining is erected, the cavity behind the lining is grouted.

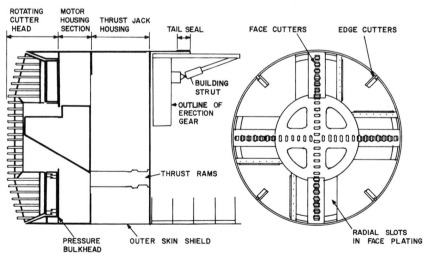

Figure 12.11 The Bentonite tunneling machine.

On the rear end of the tail a seal is fitted, which is kept in contact with the outside of the tunnel lining. The seal prevents the loss of slurry from the front of the machine. Slurry is pumped from the surface tanks to the tunnel. The slurry circuit is shown in Figure 12.12. Pumps near the tunnel face pump the slurry into the pressure chamber of the machine. The flow of slurry and the removal of excavated material out of the chamber are both automatically controlled to maintain a constant pressure at the face. Excavated material and slurry are collected in a sump within the body of the machine. It is then pumped to the surface, where the bentonite slurry is separated from the excavated material by vibrating screens and hydrocyclones and then recirculated to the tunnel. Both inflow and outflow pipes are fixed to the side of the tunnel. Telescopic pipes, which allow the pipework to extend as the machine advances, are replaced by fixed pipework when the machine is stationary during the erection and positioning of the lining.

All equipment such as the bentonite pumps, hydraulic power packs, electrical transformer, and switchgear are mounted on sleds that are

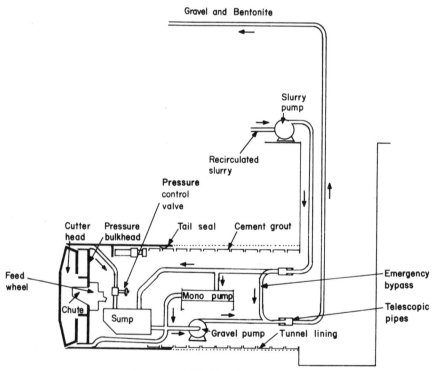

Figure 12.12 The slurry circuit.

towed behind the machine. At the beginning of a drive the machine needs to be lowered down a shaft. A timber frame can be erected behind the machine so that the initial thrust can be transferred to the opposite side of the shaft.

The Bentonite machine or hydroshield has proven to be most suitable in rock where the strength is less than 2 N/mm^2. This type of machine originated in Britain but has been extensively developed in Japan.

12.12 STABILIZATION METHODS

One method of preventing water seeping into tunnels is by the use of compressed air within the tunnel. Air locks are positioned near the beginning of the tunnel, where personnel can enter a decompression chamber. On some occasions there are two air locks so that the decompression can be done in stages. This method can be dangerous and should be managed by very experienced site personnel. There are rigid safety requirements such as medical examinations, as well as time limits for being under compression and remaining on site after being subjected to decompression. Under compression the body becomes saturated with inert dissolved gas because when gas is breathed at high pressure the excess goes into solution in the blood stream. During decompression this gas can form potentially lethal bubbles which cause the bends.

Injection processes can be utilized to change the properties of unsuitable ground such as sands or gravels by reducing the permeability or increasing the strength of the ground. The injected material can consist of cement, bentonite, or chemical grouts depending on the porosity of the ground. The grout is pumped through pipes with perforated sleeves and the ground can be treated by fan drilling from a pilot tunnel or by injection tubes driven from the surface. These methods are shown in Figure 12.13.

In a method similar to the injection processes holes can be bored at about 1.0 to 1.5 m spacing for the installation of freezing probes. Each probe is connected to a refrigeration plant, and the water bearing ground is stabilized by freezing the pore water. The use of brine or a mixture of ammonia and brine is considered to be the cheapest method of freezing, but the use of liquid nitrogen has been shown to produce a more rapid rate of freezing. The rate of freezing is very important because the major disadvantage of the method is that it can take 15 weeks or more, depending on the type of soil, to establish a frozen zone.

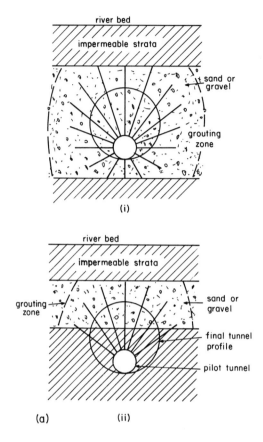

(i)

(a)   (ii)

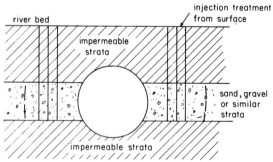

(b) Injection treatment from surface

Figure 12.13 Injection processes.

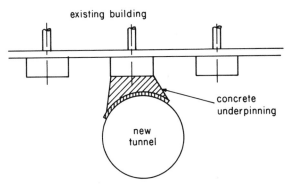

Figure 12.14 Underpinning an existing building.

If the driving of a new tunnel makes ground under an existing building unstable because the tunnel is near the foundations, it is likely that the best permanent solution is to underpin the appropriate building foundations with a concrete raft. The bottom part of the raft can be built in lean concrete through which the crown of the tunnel shield can later be driven. A sketch of this method is shown in Figure 12.14. This structural method provides rigidity and shear strength to prevent the foundations from differential settlement as the tunnel is driven below. Longitudinal prestressing is necessary on some jobs to prevent the opening of shrinkage cracks with the possible loss of shear resistance. Obviously this method can be extremely expensive and can also result in severe disruption.

12.13 MINI TUNNELS

Segmental tunneling, particularly the smooth bore flexible type, is being increasingly used for sewer and pipeline construction. Until the development of the Mini Tunnel System[37] smooth bore flexible tunnels were only possible down to a diameter of about 1400 mm, but now 1000 to 1300 mm are available.

The Mini Tunnel lining is constructed of three identical unreinforced concrete segments per tunnel ring. Longitudinal and circumferential construction joints are sealed with a bituminous strip. Each segment is erected at the tail section of the shield, which is driven forward by six hydraulic rams. The jacking load is transmitted from the segments to the shield by means of front and rear thrust rings that can be easily moved to create a larger working space for the miner. About 6% overbreak is pneumatically filled with small grade gravel through pre-formed segment grout holes. The gravel ensures immediate support to the ground as the shield advances and is then grouted to stabilize the finished work and complete the waterproofing.

12.14 IMMERSED TUBE

A feasible alternative to a bored tunnel under water is the immersed tube technique. This method basically consists of prefabricating a tunnel length, transporting the sections to site, and sinking them with ballast, if necessary, so that they rest on the sea or river bed. Advantages of the technique, compared with the more conventional methods, are as follows:

1. Any unknown geological problem in the line of the tunnel has only a very slight effect on the project

2. A shorter tunnel is possible because of the shorter gradients at each end

3. Large sections of tunnel can be prefabricated

4. Cost increases for larger diameter tunnels are not as large as for conventional tunnels

On one particular project the prefabricated section lengths were approximately 100 m long. The feasibility of the method depends on the water depth, the clearance for shipping or icebergs, the nature of the bed material, and the bed slopes.

Originally the early immersed tubes were single steel tubes, stiffened, protected, and ballasted with concrete both inside and outside of the tube. Where larger sizes are required, tunnels are now constructed in rectangular reinforced or prestressed concrete box sections.

There has been an increase in the use of the immersed tube technique over the last few years, and with further advances in technology the cable stayed semisubmersible buoyant tunnel may be possible.

12.15 PIPE JACKING

Pipe jacking[38] is a method to facilitate the construction of underground pipelines with a minimum of disruption in which concrete or steel pipes are jacked through the ground. It is an alternative to trenching when there would be interference to traffic, services, or property. The

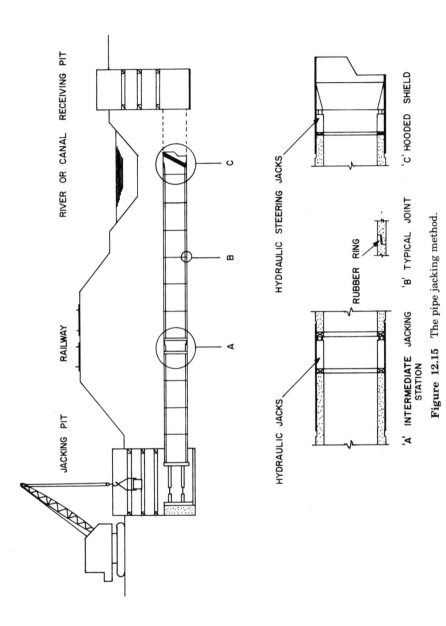

RIVER OR CANAL   RECEIVING PIT

RAILWAY

JACKING PIT

C

B

A

HYDRAULIC STEERING JACKS

HYDRAULIC JACKS

RUBBER RING

'A' INTERMEDIATE JACKING STATION

'B' TYPICAL JOINT

'C' HOODED SHIELD

Figure 12.15 The pipe jacking method.

293

method has been used for crossing under railways, canals, roads, and landing strips.

The forces required for jacking the pipes are provided by high pressure hydraulic jacks, which are interconnected hydraulically to ensure that a uniform thrust is provided, and are distributed around the circumference of the pipe being jacked by a thrust ring.

A pit is constructed to accommodate the jacking equipment. At the bottom of the pit by the outside face a thrust wall is built so that the jacks have a sturdy and stable base against which to jack. A sketch of the method is shown in Figure 12.15. The frictional resistance is reduced by the application of a suitable lubricant, such as bentonite, to the outside of the pipes, and intermediate jacking stations can be established so that the effectiveness of the jacking operation is not reduced.

To assess the practicality and economics of pipe jacking through soils or fills, factors such as the soil type and strength, ground stability at pipe level and above, water table level, and the frictional resistance between the soil and pipe should be considered. Pipe jacking is a comparatively safe method for pipeline construction and in some circumstances is a good method from an engineering point of view because the

Plate 12.1 A jumbo drilling machine. Courtesy Compair Ltd., Camborne, Cornwall, England.

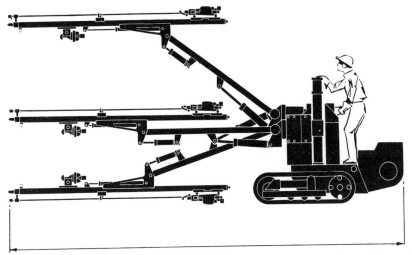

Plate 12.2 A jumbo drilling machine. Courtesy Compair Ltd., Camborne, Cornwall, England.

Plate 12.3 A tunneling shield. Courtesy W. Lawrence & Son, Basildon, Essex, England.

295

Plate 12.4 A tunneling shield. Courtesy Balfour Beatty Construction Company, Thornton Heath, Surrey, England.

pipes occupy all the volume of the excavated ground. The compaction of fill in layers, as in the trench method, is therefore obviated. Cranes, generators, compressors, pumps, pneumatic tools, grouting equipment, and rock breakers may be required as ancillary equipment.

Plate 12.5 A tunneling shield. Courtesy Markham & Co. Ltd., Chesterfield, England.

Plate 12.6 A tunneling machine for soft ground. Courtesy Sir Robert McAlpine & Sons Ltd., London, England.

Plate 12.7 View inside tailskin of tunneling machine showing motors and thrust rams. Courtesy Sir Robert McAlpine & Sons Ltd., London, England.

Plate 12.8 Heavy duty rotary excavator. Courtesy Zokor Corp., Aurora, Ill.

Plate 12.9 Tunneling machine for rock excavation. Courtesy Sir Robert McAlpine & Sons Ltd., London, England.

Plate 12.10 Rear of tunneling machine for rock excavation. Courtesy Sir Robert McAlpine & Sons Ltd., London, England.

Plate 12.11 The bentonite tunneling machine. Courtesy Edmund Nuttall Ltd., London, England.

CHAPTER 13

TEMPORARY WORKS EQUIPMENT

13.1 INTRODUCTION

A young civil engineer encounters a considerable amount of new terms when confronted with temporary works schemes for the first time on site. It is therefore appropriate to begin with an illustrated definition of terms[39].

Base plate	Usually a metal plate with a centrally positioned lug to locate a standard. The base plate transfers the vertical load from the standard to the sole plate or timber base. See Figures 13.1 and 13.2.
Bay length	The distance between the center line of two adjacent standards. See Figure 13.3.
Bearer	A timber member that can be located in a forkhead and generally forms the exterior layer of a timber grid that supports sheets of plywood for concreting operations. See Figure 13.4.
Birdcage	A three dimensional grid of scaffolding that could be used for forming the soffit of a bridge deck. The grid resembles a birdcage except for the inner part, which also contains scaffolding. See Figure 13.5.
Bracing	A diagonal system of scaffolding tubes that connect frames of scaffolding laterally. The bracing greatly reduces the possible rotation of the connection and therefore makes the scaffolding frame a continuous structure. See Figure 13.5.
Falsework	A temporary system used to support a structure during construction until it becomes safe to remove the support.

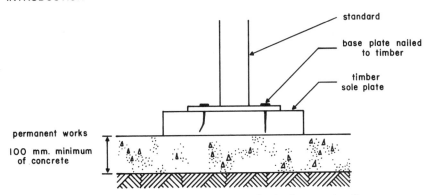

Figure 13.1   Details of base plate and sole plate on circuit.

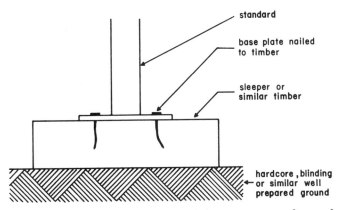

Figure 13.2   Details of base plate and sleeper on prepared ground.

Forkhead	A U shaped device in which a bearer may be located. The forkhead usually can be adjusted by means of a threaded spindle. This means of adjustment is used to accurately locate the soffit formwork to the correct elevation. See Figure 13.4.
Folding wedges	The wedging of a gap of variable thickness between two parallel surfaces by the use of two timber wedges. See Figure 13.6.
Ledger	An essentially horizontal scaffolding member.
Lift	The height of concrete cast in one pour.
Runner	See bearer.

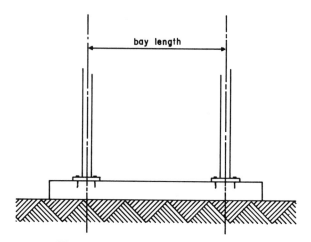

Figure 13.3 Bay length of scaffolding.

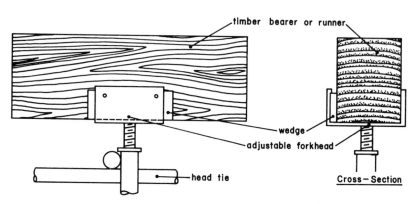

Figure 13.4 Details of bearers and forkheads.

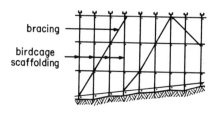

Figure 13.5 Details of birdcage scaffolding with bracing.

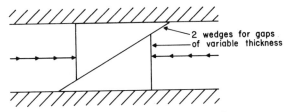

Figure 13.6 Sketch of folding wedges.

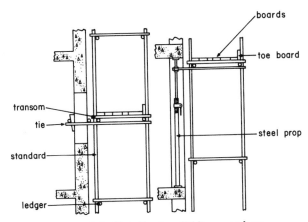

Figure 13.7 Details of scaffolding members.

Scaffolding	Any temporary elevated platform and its supporting structure consisting of tubular steel or timber members. See Figure 13.7.
Sole plate	A timber or metal member used to distribute the load from the base plate to the ground or other supporting medium.
Spigot pin	A transverse pin that is used to prevent lifts or tiers from separating.
Standard	An essentially vertical scaffolding member. See Figure 13.7.
Steel prop	An individual steel member, usually adjustable, that is used to support loads.
Tie	A scaffolding member that is attached to a rigid structure to create or improve the stability of the scaffolding. See Figure 13.7.

Toe board   An upright board secured at the edge of a walkway
                   to prevent tools, equipment, or people from drop-
                   ping off the side. See Figure 13.7.

Transom   A cross beam or essentially horizontal scaffolding
                   member. See Figure 13.7.

Wedge   A piece of timber or metal that is thick at one end
                   and tapers to a trim edge at the other end. The
                   wedge is used to adjust elevation or alignment or
                   to tighten falsework.

13.2 FALSEWORK

Falsework is a structural system that is temporarily used to support a
building or structure during construction until it becomes safe enough
to support itself, at which time the falsework may be removed. Occasionally falsework may be used on a semipermanent basis to prop up or
strengthen old structures.

The type of falsework varies considerably depending on the size and
type of job. For large and intricate jobs the falsework should be properly designed by an experienced professional engineer. The design
stage is a particularly important aspect in the use of falsework. A
disproportionate number of site accidents have occurred because of the
failure of falsework. A scaffolding subcontractor is generally employed
when large amounts of scaffolding are to be used on a project. A reputable subcontractor should be selected by the main contractor to ensure
that there are experienced designers within the organization. It is a
matter of engineering judgment whether design calculations and
drawings are necessary.

The attention of all personnel concerned with falsework design and
erection should be directed to the requirements and practices given in
the standards or codes of practice.

13.3 SCAFFOLDING PROPERTIES AND EQUIPMENT

The conventional use of standard steel tubes for shoring in Europe is
common. In North America timber props or proprietary braced scaffold
frames are generally used as shoring. A precise theoretical analysis is
complicated because of the various connections and jacking methods
when the proprietary frames are used in conjunction with timber. The

design loads should therefore be based on load tests performed by the manufacturer. Most of the steel scaffold tubes used in the United Kingdom are 48.4 mm external diameter and about 4.0 mm thick and are made from a grade of steel suitable for steel tubes used for general engineering purposes.

The dimensions and properties of conventional scaffold tubes are as follows:

Thickness of tube	4.0 mm
External diameter	48.4 mm
Internal diameter	40.3 mm
Weight	4.43 kg/m
Cross sectional area, A	567 mm^2
Moment of inertia, I	14×10^4 mm^4
Section modulus, z	5.8×10^3 mm^3
Radius of gyration	15.7 mm
Minimal yield stress	209 N/mm^2
Maximum allowable compressive stress	124 N/mm^2

From the values given above it can be seen that the minimum yield stress is a factor of $209/123 = 1.69$ greater than the maximum allowable compressive stress. The B.S.I. draft Code of Practice for Falsework[39] recommends that an overall load factor of 2.00 should be used to allow for load eccentricity and transmission through the couplers.

A reduction factor should be applied to the basic permissible stresses used in design when scaffold tube is pitted from corrosion. Very badly pitted or restraightened scaffold tubes should never be used for structural load bearing members.

13.4 DESIGN OF SCAFFOLDING

Design calculations, drawings, and specifications should accompany all falsework jobs except for the minor and nonstructural jobs. The conventional permissible stress design methods are still used in falsework design. Reference 39 states that there is insufficient information for the limit state design approach, which requires statistically significant knowledge of the load distribution and behavior of structural elements, particularly with the multiplicity of systems in use. It was therefore not possible in 1976 to recommend the general use of limit state design for falsework. However, when limit state design principles have been used in the design of permanent structures, such as reinforced concrete structures, limit state design principles may be used for the falsework that is incorporated temporarily in the structure.

The use of statistical concepts and the use of partial safety factors in limit state design affords the designer a more accurate concept of the overall structural adequacy of the falsework and is therefore a more rational approach. Limit state design should therefore be applied in falsework design as soon as the designer is confident that sufficiently accurate information has been supplied regarding stresses and loads. A scaffolding design using limit state principles is shown in Section 13.5.

In the analysis of scaffolding frames an exact analysis would be complex and would probably have little relevance in practice if the differential settlement of supports were not considered. Generally, in design a standard or vertical member is considered to be concentrically loaded by the appropriate area bounded by the middle distances between standards. A three dimensional framework is therefore not usually considered in design. If there were multiple use scaffolding layouts with the same loading, a more exact analysis might be considered worthwhile.

The standard is therefore considered as a column with fixity at each level. The exact determination of the amount and type of fixity is a complex problem. Figure 13.8 demonstrates the effect of fixity with various scaffolding configurations likely to be encountered on site and the effect the variations have on effective lengths. The definitive assignment of an effective length is difficult and is not even likely to be the same for a given configuration. The various types of fittings available on the market alter the degree of fixity. Figure 13.8 therefore attempts to demonstrate which effective length factor should be used for certain configurations. The arrows indicate which effective length factors should be assumed in design. When the arrows indicate two or three alternatives, the falsework design engineer must assume an effective length factor based on engineering judgment and experience. Usually the same factors should be prescribed for internal standards as are prescribed for similar corner standards. If a range of effective length factors is given, the larger value should generally be used for freestanding scaffolding and the smaller value used for tied scaffolding or well braced standards.

Comparisons of the various configurations shown in Figure 13.8 highlight some important aspects of falsework which are occasionally disregarded in design and practice. For example, if horizontal members such as ledgers or transoms are used at a lower level, a short distance

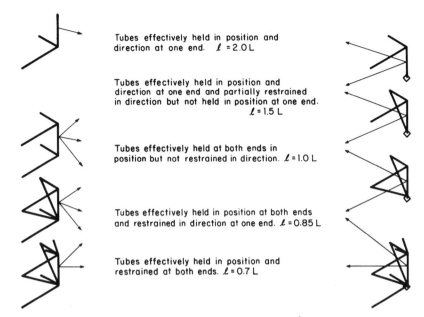

Tubes effectively held in position and direction at one end. $l = 2.0\,L$

Tubes effectively held in position and direction at one end and partially restrained in direction but not held in position at one end. $l = 1.5\,L$

Tubes effectively held at both ends in position but not restrained in direction. $l = 1.0\,L$

Tubes effectively held in position at both ends and restrained in direction at one end. $l = 0.85\,L$

Tubes effectively held in position and restrained at both ends. $l = 0.7\,L$

Figure 13.8 Amount of fixity in scaffolding tube connection.

above the base plates, the effective length of the bottom lift of standards is reduced because of the extra restraint. The extra cost of this lowest level of horizontal scaffolding may be more than offset by the extra saving in the system arising from the reduction of the effective length of the bottom lift, particularly if the design of upper lifts is based on this lift because it is the critical lift.

The permissible load on a scaffold tube depends on the slenderness ratio, the minimum yield stress, and the mode of failure. The maximum permissible ultimate load P_c may be evaluated by substituting the appropriate values in the following formula, which was adapted from the formula given in Appendix B, BS 449: Part 2: 1969:

$$P_c = A\left[\frac{Y_s + (\eta + 1)C_0}{2} - \sqrt{\left(\frac{Y_s + (\eta + 1)C_0}{2} \right)^2 - Y_s C_0} \right] \quad (13.1)$$

where P_c = the permissible ultimate load (N)
   A = the cross sectional area (mm²)
   Y_s = minimum yield stress (N/mm²)

   C_0 = Euler critical stress = $\dfrac{\pi^2 E}{(l/r)^2}$ (N/mm²)

$$\eta = 0.3 \left(\frac{l}{100r}\right)^2$$

$\dfrac{l}{r}$ = slenderness ratio = effective length/radius of gyration

The predicted bearing pressures transmitted through the framework to the ground should be compared with soil data, which should be directly established from a preliminary site investigation. If there are no records available, approximate bearing pressure values may be obtained from textbooks. The values largely depend on the type of soil. The safe bearing capacity of hard igneous rocks is about 10 MN/m^2, and capacities between 3 and 4 MN/m^2 may be assumed for the harder sedimentary and metamorphic rocks, with values between 1 and 2 MN/m^2 used for the medium hard sedimentary and metamorphic rocks. Safe bearing capacities for the softer rocks may be assumed to be as follows: compact sand and gravels and other noncohesive soils, 200 to 600 kN/m^2 except for sand, which could be considerably lower; cohesive soils, 100 to 600 kN/m^2 except for soft clays and silts, which could be considerably lower. Inspection of the soil conditions on site should determine which area in the range of values should be assumed in design.

Falsework frequently rests on fill material on site. The safe bearing pressures of fill material range from 200 kN/m^2 for broken rock to 50 kN/m^2 for firm clays provided that the fill has been fully compacted and controlled. If the fill cannot be classified, an equivalent bearing pressure of 25 kN/m^2 can be assumed unless this value is judged to be too high because of very poor site conditions. Modification factors, depending on the importance of the structure, are sometimes introduced to allow for the reliability of information on the bearing capacities. When a falsework system rests on the original ground surface, the top soil and other soft material are generally removed. If for some reason the top soil is not removed, a layer of hardcore fill should be laid above it. Timber sleepers are frequently used to distribute the load on the soil from the base plates.

The safe bearing pressure used in design is limited by the type and condition of the soil. Settlement is influenced by the nature of the loading, the dimensions of the foundation, and the degree of compaction as well as the soil type.

Diagonal bracing in scaffolding is used to prevent the rotation of connections and thereby to maintain the structural nodal positions in the frames and to provide sufficient rigidity to enable the slenderness ratio to be kept to a minimum. When horizontal forces, such as those produced by wind pressures, are applied to the scaffolding frame, the diagonal bracing transmits these forces to the ground level.

Movements arising from wind pressures acting on the scaffolding framework, particularly on the solid parts of temporary and permanent parts, should be considered. The restraining force between the foundation and the soil that prevents horizontal movement should also be carefully considered.

In areas where snow and ice are common it is very important to consider the weights of snow or ice on the horizontal surfaces and the formation of ice on the scaffolding itself if the falsework is to be used during the winter season.

13.5 EXAMPLE OF SCAFFOLDING DESIGN

Design a scaffold frame that would be suitable for erection adjacent to a 60 m high building. Three working platforms should be considered to be in use at any one time with the following loading conditions:

top platform—concreting gang	2.5 kN/m²
middle platform—bricklaying gang	2.5 kN/m²
lower platform—finishing trades	1.5 kN/m²

Assume that the safe bearing pressure of the ground, which is a stiff sandy clay, is 200 kN/m². Illustrate the design with the aid of sketches.

Answer

The scaffolding is likely to be in position from May until November, and it is unlikely that there will be any snow in the location during that period. Therefore snow loading need not be considered. The building is in a sheltered position, and the scaffolding will be tied into the building as the construction progresses upward. See Figure 13.9b. Loads from wind pressures therefore do not have to be considered in the design.

A grid for the standards of 2.0 × 1.25 m in plan will be tried in the design. Initially heights between lifts are assumed to be 2.5 m. Refer to Figure 13.9a–c. This height can be reduced to suit the tradesmen on the platform or the dimensions of the building openings, if necessary.

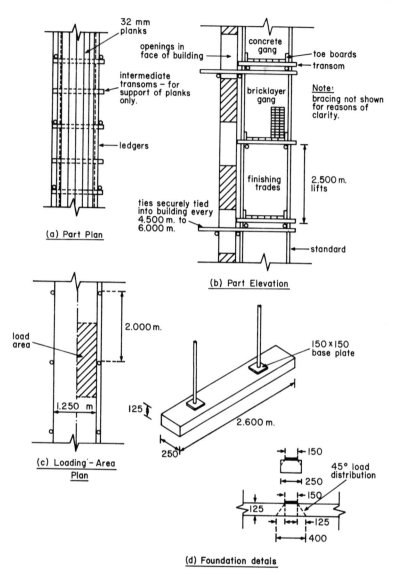

Figure 13.9 Design of scaffolding frame.

312

The maximum imposed load/standard (see loading area, Figure 13.9c) is as follows:

top platform $= 2.0 \times 1.25 \times 0.5 \times 2.5$ $= 3.13$
middle platform $= 2.0 \times 1.25 \times 0.5 \times 2.5 \times 1.5 = 4.69$
lower platform $= 2.0 \times 1.25 \times 0.5 \times 1.5$ $= 1.88$

Imposed load on standard   9.70 kN

Note that the imposed loading from the bricklaying gang on the middle platform is increased by 50% because of the possible eccentricity of the load arising from a pile of bricks.

$$\text{Imposed ultimate load/standard} = 1.6 \times 9.70$$
$$= 15.52 \text{ kN}$$

The maximum scaffolding self-weight and boards are as follows:

$$\text{No. of ledgers in 60 m height} = \frac{60}{2.5} = 24$$

Hence the dead load for the loading area considered is

Standard $=$ 60 m
Ledgers 24×2.00 $=$ 48 m
Transoms $24 \times 1.50 \times 2 \times \frac{1}{2}$ $=$ 36 m
                                    144 m

Assume that the weight of the scaffold is 44 N/m.

Dead load $= 144 \times \dfrac{44}{100}$ $=$ 6.34 kN

Allowance for diagonal bracing and boards $=$ 1.00 kN

Total maximum dead load $=$ 7.34 kN
Ultimate dead load/standard $= 1.4 \times 7.34$
                                       $= 10.28$ kN

Ultimate load/standard $= 25.80$ kN

Note that the live load should be multiplied by 1.6 or 1.5 according to which national standard the project is being designed. Similarly, a factor of 1.4 or 1.25 should be used for dead loads.

The permissible ultimate load P_c may be calculated from equation 13.1:

$$P_c = A \left[\frac{Y_s + (\eta + 1)C_0}{2} - \sqrt{\left(\frac{Y_s + (\eta + 1)C_0}{2}\right)^2 - Y_0 C_0} \right]$$

The actual length of the standard between restraints is 2500 mm. Assuming a factor of 1.0 for the effective length of the standard (refer to Figures 13.8 and 13.9) yields an effective length of 2500 mm.

$$\frac{l}{r} = \frac{2500}{15.7} = 159.24$$

$$\left(\frac{l}{r}\right)^2 = 25.36 \times 10^3 \qquad \therefore \left(\frac{l}{r}\right)^2 \times 10^{-4} = 2.536$$

$$\eta = 0.3 \left(\frac{l}{100r}\right)^2 = 0.76$$

$$C_0 = \frac{\pi^2 E}{(l/r)^2} = \frac{\pi^2 \times 200,000}{25.36 \times 10^3}$$

It is assumed that $E = 200,000$ N/mm².

Using a partial safety factor of 0.87 on the Y_s value (0.9 is used in some national standards) yields $0.87\, Y_s = 0.87 \times 209 = 181.8$ N/mm². Then

$$P_c = 560 \left[\frac{181.8 + (1.76)\, 77.84}{2} \right.$$

$$\left. - \sqrt{\left(\frac{181.8 + (1.76)\, 77.84}{2}\right)^2 - 181.8 \times 77.84} \right]$$

$$= 29.85 \text{ kN}$$

As the permissible ultimate load, 29.85 kN, is greater than the actual ultimate load, 25.80 kN, the scaffolding grid of 2.000 × 1.250 m may remain unaltered. If the values had been any closer, the designer would need to judge whether the effective length factor had been accurately estimated and whether any other factors might make it expedient to reduce the grid size or reduce the height of 2.500 m between lifts.

The dimensions of the base plate are 150 × 150 mm. Refer to Figure 13.9d for foundation details.

The safe bearing pressure of the ground for working loads is 200 kN/m².

$$\text{Working loads} = 9.70 + 7.34 \qquad\qquad = 17.04 \text{ kN}$$

$$\text{Bearing pressure from working loads} = \frac{17.04 \times 10^6}{150^2}$$

$$= 757.3 \text{ kN/m}^2$$

The ground support is therefore totally inadequate to support the stress transmitted to the base plate. Sleepers, measuring 2.60 m × 250 mm × 125 mm, are therefore used as a sole plate. The load through the sleepers is assumed to disperse at 45°. This distribution is a simplification, but the calculation is kept to a minimum by not allowing for the continuity of the sleeper from one base plate to the adjacent one. In any case the assumption is conservative.

$$\text{Bearing pressure on ground from sleeper} = \frac{17.04 \times 10^6}{250 \times 400}$$

(refer to Figure 13.9d)

$$= 170.40 \text{ kN/m}^2$$

Hence the sleepers are suitable and adequately distribute the load from the base plate to the ground.

To prevent rotation of connections, transverse diagonal bracing is placed adjacent to every alternate pair of standards.

Special proprietary quick erecting scaffolding may constrain joints so that rotation is not possible, thus obviating the transverse diagonals. Tables are also available for this type of scaffolding which give the allowable working loads for various grids and lifts.

13.6 PROPS

Props are usually made of timber or steel. The timber props can be cut to the length and end shape required, although folding wedges and packing pieces are often used for easy adjustment to the total length and also for facilitating the dismantling and removal operation. Steel props, which can be used many hundreds of times, generally contain some means of adjustment. Props are used on a civil engineering site for many functions. A few of the more common functions are shown in Figure 13.10.

To determine the spacing of props the loads can be fairly readily calculated by referring to methods shown in Chapter 11 and to a soil mechanics textbook for soil pressures. Care should be taken in calculations to ensure that an allowance has been made, if necessary, for the impact force from the discharge of a skip of concrete, a heaped pile of excess concrete, and the clustering of operatives in the concrete gang during concrete placement. If these loads can be distributed through the falsework, the anticipated dead loads can be increased by 5 and 10% to allow for the extra on site construction loads. It should be noted that the distribution of loads is less nearer the concrete face, and there-

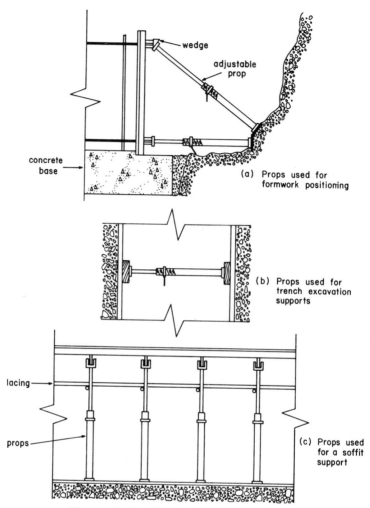

Figure 13.10 Uses of adjustable steel props.

fore the increase in load intensities due to construction forces applied
to the panels and bearers are higher than for the props.

Calculations for the prop grid size usually assume that the props fall
within the expected tolerances of 25 mm eccentricity or 1 in 40 out of
plumb. Any prop found on site to be beyond the tolerances should be
corrected immediately, and if it is felt that supervision and remedial
work cannot be guaranteed on a particular site, the safe working loads

should be reduced accordingly. Forkheads to locate the timber bearers and prevent them from being displaced should therefore be used at the top of props whenever possible. The limit to the displacement of the bearers limits the amount of load eccentricity. Lacing also helps reduce the lateral movement at the top of the prop and thereby lower its slenderness ratio. Refer to Figure 13.10c. Wedges should also be used whenever inclined props are used to ensure a perfect fit at connections. Refer to Figure 13.10a.

13.7 FAST ERECTING SCAFFOLDING SYSTEMS

There are several types of proprietary scaffolding systems that, when used, result in efficient and fast erecting and dismantling times.

A frame with one basic modular size but with variable bracing systems can be used. In conjunction with a special telescopic frame the system conforms fairly easily to the required height with a minimum of component parts. Refer to Figure 13.11.

Other types of systems include single lengths or two dimensional frames of scaffolding joined together at the connections by means of metal wedges and slotted brackets or by cups, without the use of nuts and bolts.

Tables showing maximum allowable loads are produced for the special types of scaffolding systems. These tables facilitate the design procedure. Frequently bracing of the conventional type of scaffolding is required on the special frames, but this in no way detracts from the usefulness and efficiency of some of the special scaffolding systems.

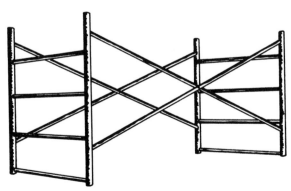

Figure 13.11   A basic proprietary scaffolding frame.

Plate 13.1 Cuplok scaffolding system. Courtesy SGB Ltd., Mitcham, Surrey, England.

Serious objections to the special systems might arise on stepped or steeply sloping terrain or very irregularly shaped soffits. Apart from these instances, however, the use of special proprietary scaffolding systems, compared with the conventional type of scaffolding, is likely to increase.

Plate 13.2 Standard frame scaffolding sections. Courtesy Gadsden Scaffold Co. Inc., Gadsden, Alabama.

CHAPTER 14

EARTHWORKS, STABILIZATION, AND PILING

14.1 EARTHWORKS

There should be a considerable amount of planning and design before the execution of a project such as the construction of a new highway where major earthmoving operations are carried out. A mass-haul diagram is usually plotted for projects requiring large volumes of cut and fill. When the excavated material is suitable for fill, the formation level should be chosen, where possible, so that there is a balance between cut and fill volumes. Bulking and shrinkage should also be taken into account.

Before the mass-haul diagrams can be plotted, a decision on the route location must be made. The most important consideration for the location of the highway is the impact on the environment. If during planning hearings the route is objected to by local residents, for good reasons, the proposed route may be realigned to a location that is very different from the most economical route from an engineering or planning point of view. Other factors such as accessibility, unsuitable fill, variability of excavated material, and the proximity of spoil tips or borrow pits may also influence the choice of route location. Better quality soils such as granular materials as opposed to cohesive types may initially cost more, but steeper permissible slopes and better compaction results may result in an overall saving. Decisions on fairly complex and wide ranging matters are therefore necessary. These decisions can only be made by a team with considerable engineering and planning experience.

Many projects include cut and fill operations. The final levels should be determined by considering the types of soil and the economics of excavating, transporting, and placing fill so that, if possible, a balance is achieved between the volume of suitable excavated material and the volume of fill required for embankments and so forth. Alternatively, it may be economically justifiable to dispose of excavated material and obtain fill from borrow pits. It is shown in Section 5.2 that cut and fill volumes cannot be directly compared and that different volumes after bulking or compaction should be considered when comparing with the in situ volumes.

Mass-haul diagrams are usually suitable for long and narrow projects, but a contour or spot elevation grid method may be more useful on a site that is wide as well as long. If it is necessary to import fill for a project, the location and choice of the borrow pit merit particular attention. A decision on the location of the borrow pit is often necessary during the tender stage. Purchase or leasing arrangements should be tentatively made dependent on the success of the bid. Frequently a soil investigation is required to establish the suitability and quantity of the source of fill under consideration. Obviously the borrow pits should be situated as close as possible to the area of fill. The sides of the borrow pit should be battered at a safe angle and the whole area should be fenced off.

Unsuitable or excess material should be disposed of in a suitable way. Conditions of disposal are usually specified at a dump.

Often, after removal of the top soil, material can be tipped on farm land and advantageously used to level off previously uneven land. After landscaping, spoil sites should be grassed over. Spoil dumps should not be placed near highways that might be widened or any other area that might be developed so as to avoid an unnecessary obstruction in the future.

In earthworks for highway schemes the gradient is another important factor to consider as well as the availability of land, environmental considerations, and the balance between cut and fill.

The tolerances permitted for the surface levels depends on which layer of the road fill is being considered. The various layers or courses that make up a highway embankment are shown in Figure 14.1. Tolerances for the surface levels, referred to in the contract documents, are usually similar to those shown in Table 14.1.

The maximum permissible number of irregularities in the flexible surface is also specified. The number of irregularities may be measured by a machine or by a straight edge.

An earthworks project may be affected immediately by a change in the weather as well as by seasonal changes. Earthworks should therefore be programmed in the most favorable seasons during the planning stage.

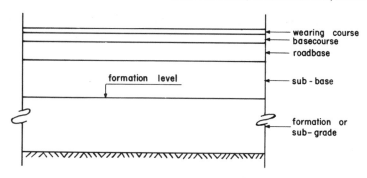

Longitudinal Section of a Highway Embankment

Figure 14.1 Courses of a highway embankment.

It is often necessary to stop work when scrapers or trucks are running over cohesive soils after a rainstorm. If the job layout merits the expense, a well laid haul road helps reduce discontinuities of work during rainy periods. Sleepers or steel mesh blankets can be useful in boggy ground under certain limited circumstances.

When a pavement is to be constructed on areas of cohesive soils and the formation is not covered immediately, a cover of undisturbed soil should be left above the formation unless a surface dressing or other temporary protection is used. This action prevents the seepage of water into the shrinkage cracks in the ground, which would cause permanent deterioration of the soil. Generally, however, very little can be done to protect earthworks from adverse weather conditions in view of the large areas involved. The overall program of the project should clearly reflect the very important constraint that the weather has on influenc-

TABLE 14.1   Surface Level Tolerances for Courses in an Embankment

Pavement	
Wearing course, flexible surface	±6 mm
Basecourse, flexible surface	±6 mm
Road base	±15 mm
Subbase	+10 mm
	−30 mm
Formation Level	+20 mm
	−30 mm

ing the scheduling of certain events. When the use of computer package programs for scheduling can be justified, great care should be exercised so that the seasonal variation on earthworks activities is properly considered. A 3 month delay in the commencement of a project can necessitate radical alterations to the logic of the program.

It has been mentioned that compaction operations may have to be stopped entirely during wet weather because of excessive amounts of moisture near the surface of the soil. Conversely, it may be necessary to add water to the soil during prolonged dry weather to achieve the optimum moisture content. Therefore soil should preferably be compacted in its final position in the fill or embankment immediately after it has been placed to avoid a deterioration in its properties due to variations in the weather.

During periods of severe frost earthworks operations tend to become more difficult and less economical. The unit weight of fill material may be considerably altered because of frost heave.

14.2 EMBANKMENTS AND OTHER AREAS OF FILL

Embankments and other areas of fill must be built up by deposits of suitable material in layers. The thickness of layers and the description of suitable material are described in the appropriate specification. Table 9.1, Section 9.3, indicates the maximum thickness of the layers of fill after compaction appropriate to the type of compaction equipment, the number of passes, and the type of suitable material. Suitable material is probably most simply explained by considering the remaining material after unsuitable material has been defined and omitted.

Material that is unsuitable for fill includes any material having a moisture content greater than the maximum specified, clays exceeding a certain liquid limit or plasticity index, frozen material, peat and other highly compressible soils, and any materials obtained from unsuitable areas such as marshes and bogs. When a combination of suitable and unsuitable material exists, the contractor is required, where practicable, to carry out the excavation in such a manner that the suitable material is excavated separately for reuse. In some circumstances the material may only be temporarily unsuitable, such as when suitable material is in a frozen condition.

Specifications or references to documents that should be consulted usually dictate to the contractor which type of compaction plant is suitable for a particular material. With the many types of compaction equipment available the contractor usually has some sort of choice for a given material. Equipment that damages or reduces the strength of the soil in its in situ state or during handling, placing, and compaction should clearly be avoided.

If the contractor wishes to remove suitable fill material from the site to suit the program of works or to use in other nearby projects, it is necessary for him to obtain the permission of the consulting engineer. Permission is usually granted if any deficit to the fill requirements is made good.

Top soil means the upper layer of earth in which plants can grow. Top soil, when present, is usually temporarily stockpiled in dumps provided by the contractor so that it can be reused for the slopes of embankments and cuttings.

Suitable material must be backfilled in areas above the natural ground level in embankments or below the formation level in cuttings where excavation has been necessary. When granular fill is required adjacent to earth retaining structures, it should consist of well graded crushed or uncrushed gravel, stone or rock, crushed concrete, or sand. It should be noted that there are restrictions on the proportion of soluble sulfate for fill material used within a certain proximity of concrete structures or cement bound materials.

The design of the embankment cross section is determined by the required width at the top, the height, and the safe angle of repose of the embankment slopes. The width of the strip of land required for the embankment should include provision for drains and manholes at the toe of the embankment. If sufficient width is not available, retaining walls may be constructed.

The consolidation and settlement of the foundation and fill material under the weight of the superimposed layers during and after construction should be carefully considered. If the embankment rests on sloping ground, the extra weight of the embankment may cause instability, and therefore a site investigation will usually be essential. When the slope is steep and particularly when the new embankment is built against an existing embankment, steps or benches in the existing slope are necessary so that keys are formed to facilitate adequate compaction of the newly placed fill by compaction equipment.

There is little chance of a shear failure occurring in the foundation material of an embankment when sands and gravels are used. The ultimate bearing capacity limits the effectiveness of clays. Silts can be difficult materials to deal with, as they can behave either as a cohesive or a frictional material. Increased strength in a silt can sometimes be obtained by building the embankment in stages, thus permitting consolidation to occur before laying another stage. This method can also be employed for other soil types, but settlement is very considerable for foundations founded on peat. Peat is therefore removed by excavation and replaced by suitable material. It can also be removed by blasting and, if the deposits are fairly fluid and can flow easily, they can be displaced by jetting.[40]

Where settlement of the embankment arising from consolidation of the foundation is acceptable, the process may be accelerated by installing vertical sand drains to enable the pore water pressure to be dissipated more quickly before the embankment is built. This method is described in more detail in Section 14.7.

14.3 CUTTINGS, PITS, AND TRENCHES

The design of the cross section of a cutting is influenced by the required width at the bottom of the cutting, the depth, and the permissible safe angle of the slopes, which in turn depends on the type of soil. The width at the bottom should be wide enough to accommodate the installation of any drainage system that might be required. The width at the top of the cutting should also be able to accommodate a drainage system, if required, so that water may be intercepted at the top and prevented from flowing into the side slopes, which would increase the likelihood of slips.

Allowance for the surface disintegration caused by weathering should be made in cuttings constructed in friable rocks. Methods that overcome the effects of surface disintegration include increasing the width of the bottom of the cutting to provide space for falling debris, reducing the angle of the side of a cutting, and the provision of berms, traps, and fencing.

The safe angle that can be specified for the slope of the side of a cutting depends on the type of soil. Nearly vertical faces can be permitted in very stable rocks, although there is usually some extra provision for a berm above certain face heights even though stability for hard rocks is independent of the height. The safe angle for some cohesive soils may be less than 20° to the horizontal.

The safe slope in cohesive soils is governed by the height of the cutting and shear strength of the soil. The stability of the slope can be determined by using a circular arc method of analysis.

If the requirements for the cutting have not been well planned and designed, considerable extra cost can be incurred both in the short and long term. Thorough site investigations are therefore necessary to ensure that the type and characteristics of the soil are known as accurately as possible. A knowledge of the presence of water and the direction of drainage and seepage is very important so that a drainage system can be planned. Drainage may be required on the slopes in the form of counterfort drains and at the top and bottom of cuttings in the form of pipes, ditches, or channels. The function of drains placed along the top of the slope is to prevent erosion and surface instability of the slope, and these should not be relied on by the designer to eliminate the chance of slips that may be caused by the seepage of water at a much greater distance within the face of the cutting. A slip can occur in a cutting when the shear strength over a fairly wide area of the soil is reduced by the ingress of water or when shear stresses are increased by additional loading at the top of the slope.

Generally the sides of pits and trenches are vertical and not battered, but they should be adequately supported by horizontal or vertical timber or metal sheeting. When persons are working in trenches, boards, walings, and struts should be used to ensure that the sides of a trench are stable. Wheel loads are likely to come very close to a trench side, and the stability of the sides is extremely important. Rigid and well erected supports must be used in most ground other than hard rock to prevent a type of accident that is unfortunately far too common on construction sites. Figure 14.2 diagrammatically shows the staging required for a trench or pit excavated by hand.

14.4 STABILIZATION

Soil stabilization may involve any method of strengthening the soil to reduce shrinkage and prevent movement. This can be achieved by drainage, the use of an impervious outer layer, or mixing the material with a more stable material so that the effect of moisture content fluctuations is not critical. Soils or aggregate can be mixed with ce-

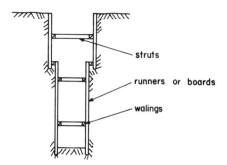

struts

runners or boards

walings

Figure 14.2 Trench support system.

ment, tar, bitumen, or asphalt. There are three types of cement bound granular materials and soil-cement. Cement stabilized materials are used for road bases and subbases to flexible pavements and also beneath concrete surface pavements, particularly for runways in airfield pavement construction. Since lean concrete increases the load carrying capacity of a pavement, the road base thickness can be reduced. It is made from washed aggregate consisting of either coarse and fine aggregate batched separately or an all-in aggregate having a range of grading, a maximum nominal size, and a water content based on the optimum for compaction. The aggregate/cement ratio usually lies between 15:1 and 20:1, that is, about 100 to 140 kg/m³ of cement.[41] Since the inception of slip-form pavers the concrete is known as wet lean concrete. Pavers give a much higher standard of surface regularity with one pass of a roller compared with the multiple passes previously used with dry lean mixes. The average 28 day strengths should range between 8 and 20 N/mm². Lean concrete is used for road bases or subbases for most types of traffic intensity.

The use of cement bound granular materials is a form of stabilization limited to subbases for all types of road or road bases with a medium traffic intensity. Granular materials may consist of natural gravels, washed crushed rock, or slag having a specified grading, water content, maximum permissible sulfate percentage, and a maximum permissible percentage of material in the silt clay particle range. The cement content, usually between 4 and 8%, say, 85 to 170 kg/m³, is chosen to give a minimum average 7 day strength of about 3.0 N/mm². The granular material and cement should be mixed in a paddle or pan type mixer of the forced action batch or continuous mixer.

Soil-cement construction is common for small airfields and is quite common for subbases and roads in certain parts of Africa and North America where traffic intensity is light. The materials used range from soil, chalk, pulverized fuel ash, a washed or processed granular material, crushed rock or slag, well burnt shale, to any combination of these materials. The specification limits the sulfate content to a certain maximum percentage, about 1%, states the grading limits, and requires a 7 day strength of about 3.0 N/mm² using between 80 and 250 kg/m³ cement content. When stationary plant is used, it should be of the power driven paddle or pan type and may be of the batch or continuous type. If batch mixers are used, the appropriate amounts of material, cement, and water must be measured and special care must be taken to ensure that the cement is spread uniformly in the mixer for thorough mixing. The paddles, baffles, water content, and rate of feed must be adjusted accurately when continuous mixing equipment is used to ensure that the material is uniformly mixed. The soil-cement mixture is limited to subbases and road bases with relatively light traffic intensities.

For the road base, base course, and wearing course, aggregate of various ranges of grading can be used mixed with tarmacadam, bitumen, or asphalt.

Any layer of cement stabilized material that is not covered by another layer within a few hours normally requires curing for about 7 days after compaction. The material can be cured using a specified layer of suitable soil, impermeable plastic sheeting, bituminous spraying, or other specified methods.

Another type of soil stabilization has been developed in France. This method,[42] known as reinforced earth, consists of earth, reinforcement, and skin elements. Soils that range within a grain size greater than cohesive soil and smaller than large lumps of rock are suitable.

The reinforcement can be metals, plastics, or timber in the shape of strips, wires, meshes, or bars. It improves the mechanical properties of the soil and provides internal stability within the soil. At the free boundaries of the earth structures it is necessary to provide a skin to prevent the earth from flowing away between the reinforcement.

14.5 PAVING MACHINES

Transport vehicles carrying plant-mixed material should be capable of discharging cleanly and should be protected from the weather during transit. There should be a minimum of delay in the loading, transportation, tipping, and spreading operations. Road base material is now usually spread by a paving machine. The equipment usually requires approval before it can be used initially. It should cut slightly into the previously laid lane edge to ensure that the material has been properly compacted. As an alternative to a paving machine a spreader box may be used.

Compaction of the material should be completed as soon as possible after the material has been spread. If it is not done before a certain specified time, say within 2 hours, the specification may require cement stabilized material to be broken out and relaid. Special attention should be paid to the compaction of the material at the joints by using small compactors if necessary. If any layer of cement stabilized material is not covered within a few hours, the protection and curing process, mentioned in Section 14.4, should be used.

Slipform paving machines are able to place a concrete road slab

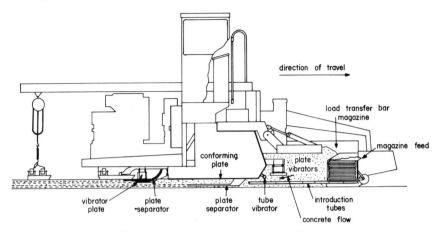

Figure 14.3 A slipform paving machine.

in one single operation. If reinforcement is required, fabric reinforcement previously laid on the formation can be lifted by a prelifter and incorporated in the concrete slab. The concrete is generally fed into a front hopper or magazine, using ready mixed concrete trucks. As the concrete flows through the magazine, it is vibrated by plates. Conforming plates then set the in situ concrete to the required level, and the surface of the slab is finally vibrated before the machine travels continuously on without interruption. Refer to Figure 14.3.

The machines[43] can be self-leveling, using a wire or kerb as a reference, and can automatically insert dowel bars for transverse and longitudinal joints.

A similar machine can be used for laying most black base and wearing course materials. In fact some machines are equipped for multiple use with little alteration to the basic machine. The machines can be equipped with either a tamping screed or a tamping/vibrating screed and are suitable for laying crushed stone, dry lean concrete, and most black base and wearing courses.

Black base or wearing course material is transported by truck to the construction site. On arrival the driver positions the truck so that it can be pushed by the paving machine. The truck then proceeds to tip the material into the hopper. The hopper can sometimes be hydraulically raised and can also include folding sides to direct the material onto the bar-conveyor system. The material is conveyed to the auger box through power operated flow gates. When the auger receives the material from the conveyor, it spreads the material evenly on the course below, in front of the leading edge and across the complete width of the screed. Finally, the sole plate of the machine ensures that the material is compacted to the required depth and width.

14.6 PUMPING AND DEWATERING

Construction in waterlogged ground should be avoided in the design stage for the foundations of a structure, if at all possible. The design engineer should consider the use of shallow raft foundations so that excavation and construction can take place above the groundwater level, or alternatively, consider the use of piled foundations so that the loads can be transferred to underlying firmer strata, with the pile caps in a position above the groundwater level. In some circumstances it may be possible to avoid waterlogged ground by relocating structures during the planning stage, particularly on large jobs where several structures cover a large area and where the ground is waterlogged over a comparatively small area.

Sometimes, however, it is not possible either to avoid construction work below the groundwater table or to show that the avoidance of construction work below groundwater level is economically justifiable. In these cases the cheapest method of keeping the excavation free of water is by pumping, unless the sides of the excavation are unstable because of soil conditions, in which case it is preferable to remove the water before it flows into the excavation.

An important factor in the planning of a dewatering method is the permeability of the soil. Darcy's law may be applied for most conditions of groundwater flow where:

$$Q = Aki \qquad\qquad (14.1)$$

and

Q = rate of water flow (m³/s)
A = area through which flow occurs (m²)
k = soil permeability (m/s)
i = hydraulic gradient = h/r, where h is the height above datum over a radial distance r

The permeability k may vary from 1.0 m/s in clean gravelly soils to 1.0 μm/s for very fine sands and clay/silts and onto the virtually impervious clays, with a k value of about 1.0 nm/s (n = nano = 10^{-9}).

The determination of the soil permeability is important in making any large scale dewatering analysis. The most reliable method is considered to be the use of full-scale pumping tests from a well point within the site area. Other methods of determining the soil permeability include rising and falling head tests in a borehole, computing values from flow conditions measured by piezometers in boreholes, computing values from soil properties, and by comparisons with previous work, if any, in the same area. For large jobs where permeability appears to be an important factor, particularly in coastal or esturine areas, the advice of an engineering geologist should be sought. Where the groundwater table is likely to be below the excavation level, there is generally little or no work done for the determination of the soil permeability although the effects of the surface runoff of water should always be considered. The contractor is normally required to arrange for the rapid dispersal of water in the specification.

The rate of flow of water to be removed from an excavation depends on the precautions taken to exclude water, the type of soil, and the head causing the water flow. It has been customary over many years to pump the water from pumps within the main excavation. This method is shown in Figure 14.4. It is a method that is easy to install, easy to use, and flexible. Furthermore it is a very common method and is successful where the stability of the soil is not critically impaired by the action of seepage forces through the face of the excavation.

There are several types of pumps used frequently on construction sites:

1. Reciprocating
2. Diaphragm
3. Ejectors operated by compressed air
4. Centrifugal

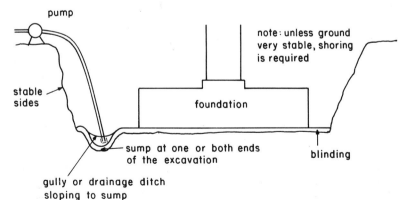

Figure 14.4 Cross section of a typical surface pumping method.

Reciprocating pumps are mainly used for pumping fairly clear water. A piston operating in a cylinder sucks in water through the inlet which is then forced out through the outlet. Gasoline run diaphragm pumps, which operate by the oscillation of a flexible diaphragm, are generally capable of being used in muddy water and of withstanding fairly rough usage on site. Ejector pumps consist of chambers that are filled with water which is then ejected by using compressed air. Centrifugal pumps contain vanes or blades that rotate on a vertical or horizontal spindle. These pumps are versatile, easy to maintain, and capable of operating in muddy water.

Pumps are required for expelling water from excavations, pits, and tunnels and for use in dewatering methods such as well points and cofferdams. Often the greatest pumping capability is required at the beginning of the job. For example, pumping becomes a lot easier when the sheet piling interlocks tighten, after a differential head of water has been established through the initial pumping. When pumps are used for long periods, a slow running oversized type of pump is preferable to reduce wear and maintenance.

Submersible pumps available from the manufacturers have a typical range in performance which suits most conditions. Submersible pumps are available which pump 1000 liters/min at 3.0 m head to 300 liters/min at 18.0 m head. Sludge pumps are designed for low heads with a capacity of 200 liters/min at 3.0 m head to 100 liters/min at 30.0 m head.

Alternative methods to combat the presence of water include freezing and electro-osmosis. The electro-osmosis method, which is particularly successful in uniform beds of fine silt, uses an electrical potential to drive the water to negative electrodes at the wells, and uses expendable metal rods as the positive electrodes. Unstable ground can also be frozen, using liquid nitrogen, where excavations or tunneling operations take place. Other methods include the use and application of sand drains and well points. These are discussed in the following sections.

14.7 SAND DRAINS

The naturally slow process of the consolidation of compressible saturated soils such as clays, silts, and other alluvial deposits can be accelerated by improving the drainage within the soils. Before the construction of an embankment, for instance, vertical sand drains can be installed in the foundation material to accelerate drainage. The purpose of the vertical sand drains is to enable the pore water pressure to be dissipated more quickly, thus increasing the shear strength of the soil.

If some soils are loaded by an embankment or a temporary surcharge fill, the water pressure in the foundation soil increases. A dangerously unstable situation can arise if the water is unable to escape sufficiently quickly. This situation can be relieved by controlling the rate in the formation of the embankment or by the installation of sand drains. By introducing the sand drains, which are vertical columns of sand, the drainage path of the pore water is considerably decreased. The drainage time is considered to be roughly proportional to the square of the length of the drainage path, and therefore the provision of sand drains considerably accelerates the water dissipation.

Sometimes the embankment is built higher than the final elevation. This overloading with a surcharge is used on soft deposits to bring about the desired settlement more rapidly. The surcharge is then removed from the top of the embankment when the calculated settlement has been achieved and can then be reused on another embankment in the same project, if required and if compatible with the program.

Sand drains are installed by various methods. On small projects the vertical boreholes can be formed using bored pile equipment with a casing to line the hole. For large schemes the columns can be formed by driving a hollow steel tube fitted with an expendable shoe or a flat hinged bottom plate. As the tube or mandrel is withdrawn, the space is backfilled with sand. Sometimes the hole may be formed by jetting. The rate of withdrawal and the applied pressure on the sand must be carefully regulated. Problems associated with the installation of sand drains may be reduced by the use of sandwicks.[44] Sand is placed in a water permeable stocking with a diameter of approximately 65 mm and a length that is made to suit the depth of the compressible layer. The stocking is then lowered into the hole, which is formed by drilling or driving a tube or by boring. The continuity of the sand drain is assured by using the sandwick system, and there is less of a tendency for the sand column to act as a pile.

The unit weight of the sand in the stocking can easily be varied according to the design requirements. The sand filled stocking also prevents any arching effect in the sand, which is sometimes a problem when the sand is backfilled into the hole in the normal way.

The rate of loading of an embankment should be controlled to ensure that the soil does not fail and the sand drains do not fail in shear. The measurement devices for controlling the rate of settlement and measuring the pore water pressure at various depths are the settle-

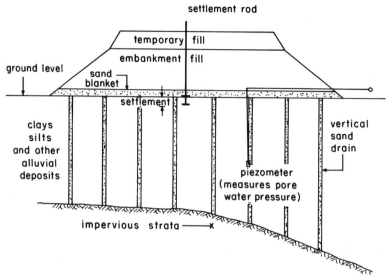

Figure 14.5 Vertical sand drain.

ment observation rods and piezometers, which are shown diagram-matically in Figure 14.5.

After the installation of the sand drains, a blanket of free draining sand or granular material is spread over the whole drain area so that the water can dissipate from the drained soil.

The depth and spacing of the sand drains depend on the depth to an impervious stratum and the type, condition, and permeability of the soil. The spacing can vary between about 0.8 and 4.0 m, with depths between 3 m to 20 m. The required number of sandwicks is greater than the in situ systems of larger diameter drains, and therefore greater coverage is provided.

The disturbance of the soil caused by the installation technique may considerably influence the performance of the sand drains. Sand drains have usually been ineffective when used in peat.

14.8 GROUNDWATER LOWERING SYSTEMS

A well point is a mechanical device that can be installed in the ground and withdrawn. It consists of a large pipe, generally about 50 mm diameter, with a perforated screen at the foot. A two valve apparatus is located within the perforated screen. During the jetting stage a ball in the ball valve remains in the open position and is kept equidistant from the well point casing by a retaining cage. This ensures that a positive streamline flow of water through the tip, giving maximum jetting pressure, is created to give an efficient erosive action for fast installation. During the dewatering phase the ball valve closes and prevents the entry of fine particles while a ring valve, which is open, only permits water to flow freely through the screen. The perforated length is covered with a fine metal gauze to prevent entry of fine particles of soil into the well point.

When the well point is being installed by jetting, the water pressure and flow are reduced as it nears its final depth, and coarse sand can be poured into the annular space created around the well point.[45] This acts as an outer filter around the mesh and provides a vertical drain that facilitates drainage so that the groundwater level is lowered as efficiently as possible.

The spacing of the well points depends on the permeability and type of soil, varying between 0.5 and 1.5 m. Because of the flow of water the more permeable rocks require a closer spacing.

Since the depth of a single system is limited to about 7.0 m, deep excavations require two or multistage systems. Refer to Figure 14.6.

Although initial installation costs are relatively high, the rapid jetting permits quick and easy installation and the excavation costs are substantially reduced in dry soil conditions because the amount of shoring is reduced or eliminated and the work below ground can proceed more easily and efficiently.

Shallow wells, formed by boring methods, with diameters of 300 mm or more are sometimes used with filter tubes of 150 mm diameter. This system has the advantages that obstructions in the ground can be removed during the boring operation and that the ground conditions can be examined over the total depth so that modifications to the well spacing or filter system can be made. The installation of this type of well for a multiwell system is more costly than the jetting well point system and takes a longer time to install.

14.9 PILING

Piles are basically columns in the ground which transfer heavy foundation loads to the lower levels of the soil or rock strata. They are generally long and slender, and the cross sectional shape, although usually circular or square, can possess various geometrical configurations. Steel, reinforced concrete, precast concrete, steel concrete com-

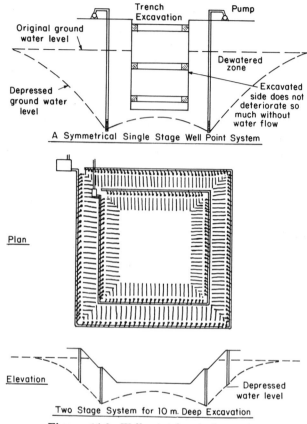

Figure 14.6   Well point dewatering system.

binations, and timber are frequently used for piles. Installation methods include piles that are driven by the blows of a hammer, bored cast-in-place piles, driven tube cast-in-place piles, screw piles, jacked piles, and piles that are installed by a combination of several of the above mentioned methods.

There are two ways in which a pile can transfer the load from the foundation into the soil. An end bearing pile transfers most of the load through a fairly soft soil to a hard rock or very dense soil at the lower level, and is capable of withstanding the working load without any significant settlement. Friction piles rely mainly on the skin friction between the soil and the face of the pile for their load transfer. Friction occurs when a pile does not reach an impenetrable or compact stratum and remains for all of its length in soils such as clays or silts.

When solid or closed ended tubular piles are driven, the surrounding soil is displaced by fairly large amounts as the pile is driven into the ground, whereas a cylindrical column of soil is removed in a bored pile operation and therefore no soil is displaced laterally. The difference in the large displacement and nondisplacement categories is important in calculations to determine the load carrying capacity of the piles. When hollow open ended sections or piles with narrow elements are driven, only a relatively small displacement occurs during driving.

To increase the end bearing capacity of a pile the base can be enlarged in driven in situ piles as shown in Figure 14.7a, or in bored piles an underreamed bell mouth can be formed, as shown in Figure 14.7b by opening the drill out at the bottom of the bored pile. The enlarged base is formed by holding the tube stationary with heavy cables when it has been driven to a suitable depth, and then by continuous hammering and successive charges of concrete, the concrete is forced out laterally to form the enlarged base.

A pile must be designed to be capable of resisting the internal stresses when subjected to the worst combination of all the possible loads and limit states. It must also be capable of withstanding the stresses imposed on it during installation. In addition to the design calculations for the pile, calculations must also show that the soil is capable of withstanding the applied loads. The design of the pile from the structural point of view, which takes the form of a fairly typical analysis, can be found in various textbooks concerned with foundation design.

If a pile is preformed, it must be capable of resisting all the stresses imposed on it during handling and installation. During the installation of a driven pile, failure can occur at the head of the pile because of

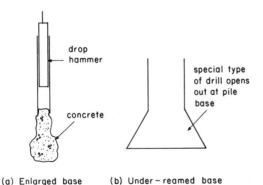

(a) Enlarged base   (b) Under-reamed base

Figure 14.7 Methods of increasing end bearing capacity.

local failure caused by a malfunctioning helmet, insufficient area of pile material, or misalignment between the pile head and the direction of the hammer blow. It should be noted that if piles are driven through an unexpectedly stiff stratum, excessive driving stresses can result because design assumptions are exceeded.

The ultimate bearing capacity of a pile depends on the size, cross sectional shape and type of pile, and the properties of the soils in which it is embedded. It can be approximately determined by means of a loading test, use of a pile formula based on dynamics, or use of a pile formula based on statics and a soil mechanics analysis of the soil.

If a loading test is conducted on a particular pile, the ultimate bearing capacity can be determined. However, there is difficulty in precisely defining the ultimate bearing capacity, and for this reason a load applied at the head of the pile, which causes a settlement of 10% of the pile diameter, is generally regarded as the ultimate bearing capacity. Soil samples should be collected from boreholes intended for bored piles, and rates of penetration should be measured for driven piles to assess the uniformity of site conditions.

A very approximate value of the ultimate bearing capacity of a pile may be determined by using a dynamic pile formula in which the driving characteristics of each individual pile are used. Driving formulas are not directly applicable to piles whose toes rest in cohesive soils, and are inapplicable to some types of pile driving equipment. The use of dynamic formulas is therefore little used today. However, measurements of the driving resistance, previously associated with the dynamic formulas, are still extremely useful measurements to record for every pile, in the form of the number of blows for a given depth of penetration.

The ultimate bearing capacity of a pile, including its own weight, is supported by the skin frictional resistance between the surface area of the pile and the soil and the end bearing resistance at the base of the pile. Usually it is assumed that both resistances reach their ultimate values before failure occurs. This relationship can be expressed by the following equation:

$$Q_u = Q + W = fA_s + A_b(q + P_0) \qquad (14.2)$$

where Q_u = ultimate failure load on the pile
Q = ultimate load applied to pile
W = weight of the pile
f = average skin friction per unit area
A_s = effective surface area of embedded pile shaft
A_b = plan area of the base

q = ultimate value of the resistance per unit area of the base due to the shear strength of the soil

P_0 = the overburden pressure at the level of the base

14.10 SHEET PILING

Steel sheet piling is used in many types of temporary and permanent civil engineering works. It is frequently used as a temporary means of preventing the seepage of water into the works and also of supporting the surrounding ground. Sheet piling is also used in docks and harbor works, canal and river bank strengthening, and seepage prevention works, sea defense works, caissons, and other permanent foundation works.

Sheet steel sections are designed to provide maximum strength and durability to withstand the stresses imposed by the earth and water pressures and the stresses induced during the driving and extracting operations. As the stresses vary for each site location, the sheet piles are manufactured in a complete range of weights, thicknesses, and sizes.

The steel sections are made so that each edge interlocks with the next to facilitate the pitching, driving, and extracting operations and also to create an effective water seal once a difference in water level has been established on each side of the sheet piles. In some sheet piling systems the piles are very often pitched, driven, and extracted in pairs (see Figure 14.8). Special corner piles are supplied, and a range of acute and obtuse corners for various angles is available. Alternatively, if a specific dimension is required, it may be necessary to weld and fabricate an individual nonstandard section. Junction piles, which are also available, normally consist of a half pile or part pile intermittently welded to an ordinary pile.

The maximum length of sheet piling for each site location depends on the type of soil strata and water level encountered, the penetration required, and the type of construction for which the piling is required.

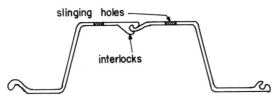

Figure 14.8 A pair of steel sheet piles.

The maximum length of the heaviest types of section is approximately 25 m, although the maximum free height for an unbraced cantilever section can be very much smaller even for the heaviest section. Some system of bracing is therefore required.

Sheet piles should be guided into position with guides that are held rigidly in position at two different levels. The guides and guide frame should be substantial and robust. The guides often consist of 300 × 300 mm timber, and the distance between the guides should be about 6 mm greater than the overall pile depth so that packing pieces or wedges can be inserted to maintain a vertical line of piles in one direction. In the other direction there is a tendency for sheet piling to lean forward in the direction of driving. A good driving procedure should generally prevent any serious leaning, but if a lean appears to be developing, a rope attached to a winch may be used to correct the leaning tendency or to ensure that there is no serious departure from the vertical.

A sheet pile driving system is shown in Figure 14.9. A competent, experienced sheet piling foreman is necessary to ensure that the job proceeds on time and that problems are kept to a minimum. Most foremen use their own preferred procedure, but these do not differ essentially from the following procedure.

About 6 to 12 pairs of piles are pitched and interlocked into position to form a new panel of sheet piles, after the first pair of piles is pitched and partly driven. The last pair of piles is then partly driven. This stage in the procedure is close to the stage depicted in Figure 14.9. Special care should be exercised to ensure that the first and last pair of sheet piles remain perfectly vertical as the two end pairs guide the other pairs in the panel. The width of the panel may be controlled by increasing or decreasing the width of each pair of piles by slightly tilting the piles in plan section. The remaining piles are then driven to an elevation that permits the piling hammer to operate at an elevation immediately above the top set of guide wales or to their final elevation, depending on the height of the piles. This sequence is then repeated around the entire perimeter. The wales positions of a typical timber pile frame are shown in Figure 14.10.

A double acting hammer suspended from the jib of a crane and fitted on a pair of piles can be used for driving. The selection of the type and size of the hammer to be used for a particular job depends on the properties of the soil, the depth of penetration, and the type and weight of pile. A light fast hammer should be employed in difficult ground conditions where boulders or other obstructions are expected. Vibro-hammers, with frequencies ranging from 10 to 20 cycles per second,

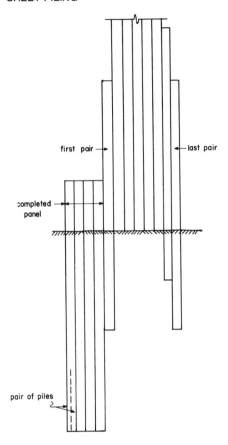

first pair

last pair

completed
panel

pair of piles

Figure 14.9 Sheet pile driving sequence.

are now often used because of their relatively quiet operation in popu-
lated areas.

Single skin sheet piling should be designed to withstand the water
and earth pressures on the outside of the piling. In tidal areas it may be
necessary to design the piling to withstand water pressure on the inside of the piling.

A cantilever retaining wall is designed by considering the active and
passive pressures that develop, as shown in Figure 14.11. Note that the
passive pressure P_{P2} is assumed to develop at the bottom of the piling
and on the same side as the active pressure so that the end fixity of the
cantilever can be created. The depth of penetration of a cantilever pile
should therefore be sufficient to prevent forward movement and rotation of the pile. When the overall depths are large, this penetration is

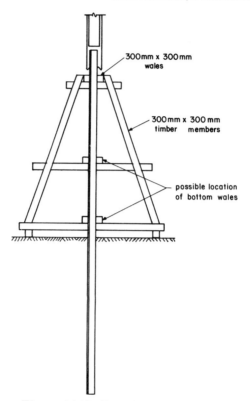

Figure 14.10   Typical sheet pile frame.

not sufficient to prevent rotation at the top and therefore the top of the piling must be anchored by a tie rod. Anchored retaining walls can be designed in two ways. One method assumes that the passive pressure is not sufficient to prevent rotation and that the design concept for the sheet piling is similar to a simply supported beam. The alternative method assumes that the passive pressure in front of the piles is sufficient to prevent rotation of the toe of the pile so that this end of the piling may be assumed to have a fixed boundary condition. The piling is then designed by a method similar to a propped cantilever analysis. Obviously the bending moments can be substantially reduced by the use of tie rod anchorages, and consequently the cross sectional weight of the sheet piling can be less. This saving, however, must be offset against the cost of providing tie rods and anchorages.

The sheet piling should extend sufficiently below final excavation level to reduce pumping to a minimum and to ensure stability by

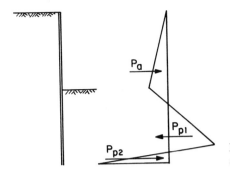

Figure 14.11 Pressures on a cantilever sheet piling wall.

preventing the soil from being forced into the excavation from underneath the toe of the pile by the external soil and water pressures.

14.11 COFFERDAMS AND CAISSONS

A cofferdam is a temporary structure built around the perimeter of a foundation for the purposes of excluding water and supporting the surrounding soil to permit the construction of the works without the need for excessive pumping operations. Often a cofferdam consists of a single line of sheet piling. A cellular cofferdam consists of a series of filled cells without ties, and with a circular or curved shape in plan. Two types, the circular type and the diaphragm type, are fairly common and are diagrammatically shown in Figure 14.12. The double wall

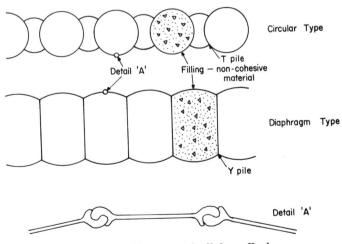

Figure 14.12 Two types of cellular cofferdams.

cofferdam consists of two parallel lines of sheet piling tied together and filled in between with a cohesionless fill such as sand, gravel, or hardcore. The tie rods secure the two lines of sheet piling at the top and, when possible, at a position about halfway down to the excavated level.

A caisson is very similar to a cofferdam. It is a structure that is sunk through soil and water for the purpose of excavating and construction of the foundation within which it is permanently incorporated.

The first activity in the sequence of construction of a cofferdam is usually to dredge the foundation if the foundation is to be constructed in a river, lake, or the sea. This dredging can be done with a grab or clamshell bucket. If piling is necessary, it can be installed after the dredging or after the installation of the sheet piling. At this point the bracing frame can be installed. The sheet piling is then driven, and a concrete seal is poured through a tremie pipe to the base of the foundation. After this the water is pumped out. Considerable pumping may be necessary initially to establish a difference in head between the inside and outside of the cofferdam so that the interlocks tighten unless use can be made of the variation in the high and low water levels. If, after a

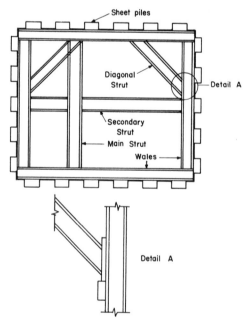

Figure 14.13 A plan of a cofferdam.

difference in head has been established and water still leaks in at an undesirable rate, ashes or other fine material may be scattered outside the cofferdam to improve the seal of the interlocks. After the foundation structure is built to above high water level, the cofferdam is removed after flooding.

Generally piles are withdrawn by the use of a power driven extracting hammer or vibrating extractor. Hydraulic jacks can also be used to extract stubborn piles where appropriate.

Cofferdams can be braced by one or more levels of wales, main, and secondary struts which can consist of steel, timber, or reinforced concrete. If prefabricated bracing is used, allowances should be made for the deflection of the piling as excavation proceeds. It may be possible to replace some of the parallel main and secondary struts by diagonal struts to create more room for grab buckets or concrete skips. Refer to Figure 14.13.

14.12 WORKED EXAMPLES

Example 14.1

A pump in a well is pumping water through sand to a depth of 25 m. Determine the rate of water flow if the drawdowns, that is, the drops in the water level, are 2.5 m at 10 m from the well and 0.5 m at 25 m from the well after a steady state has been reached. Assume that the soil permeability of the sand k is 5×10^{-5} m/sec.

Answer 14.1

$$Q = Aki \qquad \text{(see equation 14.1)}$$

$$= 2\pi r h k \frac{dh}{dr}$$

where h = height of water above base level
r = radius from well

Integrating yields

$$q \int_{r_1}^{r_2} \frac{dr}{r} = 2\pi k \int_{h_1}^{h_2} h\, dh$$

$$q = \frac{\pi k(h_2^2 - h_1^2)}{\log_e \frac{r_2}{r_1}}$$

$$h_1 = 25 - 2.5 = 22.5$$
$$h_2 = 25 - 0.5 = 24.5$$
$$r_1 = 10.0$$
$$r_2 = 25.0$$

$$\log_e \frac{25}{10} = 0.91629$$

$$q = \frac{\pi \times 5 \times 10^{-5} \times 47 \times 2}{0.91629}$$

$$= 0.0161 \text{ m}^3/\text{sec}$$
$$= 967 \text{ liters/min}$$

Example 14.2

Determine the ultimate load that may be applied to a concrete pile in a sand where the average skin friction is 50 kN/m^2 and the sum of the ultimate value of the resistance per unit area of the base due to the shear strength of the soil and the overburden pressure at the level of the base may be assumed to be 10,000 kN/m^2. The pile is 300 mm in diameter and has an effective depth of 12 m.

Answer 14.2

The ultimate load applied to the pile may be determined from equation 14.2:

$$Q + W = fA_s + A_b(q + P_0)$$

The skin friction resistance is

$$fA_s = 50 \times 12 \times \pi \times 0.300$$
$$= 565 \text{ kN}$$

The base resistance is

$$A_b(q + P_0) = \frac{\pi}{4} \times 0.300^2 \times 10,000$$

$$= 707 \text{ kN}$$

The weight of the pile is

$$W = \frac{\pi}{4} \times 0.300^2 \times 12 \times 23.6$$

$$= 20 \text{ kN}$$

Therefore the ultimate load that may be applied to the pile is

$$Q = 565 + 707 - 20$$

$$= 1252 \text{ kN}$$

Note that this example has been deliberately simplified compared with the usual ultimate load calculations for piles in practice. Properties vary for the same type of soil at different depths and, of course, for different strata along the length of the pile. The assessment of the values of the properties of the soil in practice is often difficult.

Example 14.3

Sketch the net pressure diagram and the mode of deflection for the following single sheet steel pile retaining walls:

(a) A cantilever wall
(b) An anchored wall where the passive pressure is sufficient to resist the forward movement of the toes of the piles but rotation occurs
(c) An anchored wall where the passive pressures are sufficient to restrain forward movement and rotation of the toe

Answer 14.3

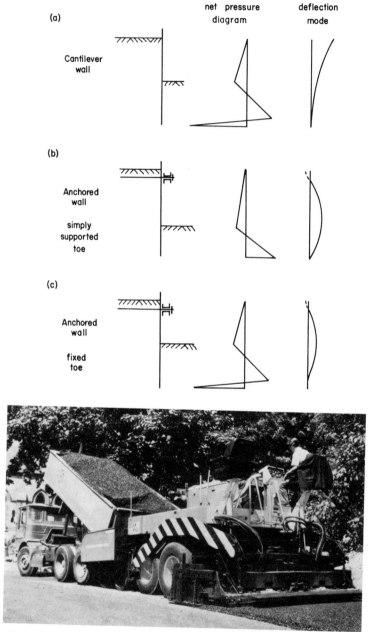

(a)

Cantilever
wall

net pressure
diagram

deflection
mode

(b)

Anchored
wall

simply
supported
toe

(c)

Anchored
wall

fixed
toe

Plate 14.1   A paving machine. Courtesy Blaw Knox Ltd., Rochester, Kent, England.

Plate 14.2 Sandwick installation. Courtesy Cementation Ground Eng. Ltd., Rickmansworth, Herts, England.

Plate 14.3   Sheet piling equipment. Courtesy T. Smith & Sons Ltd., Leeds, W. Yorks, England.

REFERENCES

1. A. J. Christian, *Proceedings of CIB W-65 Second Symposium on Organization and Management of Construction*, V-39, and Discussion, Session V, Haifa, Israel October–November 1978.
2. A. Casagrande, A Classification and Identification of Soils, *Proceedings of the American Society of Civil Engineers*, New York, Vol. 73, 1947, p. 73.
3. R. De Neufville, E. N. Hani, and Y. Lesage, Bidding Models, *Journal of Construction Division*, P.A.S.C.E., New York, March 1977.
4. *Civil Engineering Standard Method of Measurement*, I.C.E., London, 1976.
5. *The I.C.E. Conditions of Contract*, 5th ed., I.C.E., London, 1973.
6. Structural Safety, *Canadian Building Digest*, CBD 147, N.R.C., Ottawa, 1972.
7. H. N. Ahuja, *Construction Performance Control by Networks*, Wiley, New York, 1976.
8. *An Introduction to Engineering Economics*, I.C.E., London, 1969.
9. B.S. 1757:1964, Power Driven Mobile Cranes.
10. U.S.A.S. B 305:1968, Crawler, Locomotive and Truck Cranes.
11. B.S. C.P. 3010:1972, Safe Use of Cranes.
12. *Fundamentals of Earthmoving*, Caterpillar Tractor Co., Peoria, Ill.
13. *Caterpillar Performance Handbook*, 9th ed., Caterpillar Tractor Co., U.S.A., 1978.
14. *Specification for Road and Bridge Works*, HMSO, 1976.
15. A. F. Toombs, The Performance of Vibrating Rollers in the Compaction of Soils. TRRL Report LR 480. Dept. of the Environment, Berkshire, England, 1972.
16. B.S. 1377:1975, Methods of Testing Soils for Civil Engineering Purposes.
17. ASTM D 698:1970, Moisture-Density Relations of Soils Using 2.5 kg. Hammer.
18. ASTM D 1557:1970, Moisture-Density Relations of Soils Using 4.5 kg. Hammer.
19. *Guide to Hyster Compaction*, Hyster Co. Publications, Kewanee, Ill.
20. CIRIA Research Rept. No. 1, London.
21. *Concrete Pressure on Formwork*, CIRIA, London, 1979.
22. *Formwork for Concrete*, SP No. 4, 3rd ed., American Concrete Institute, Detroit, 1973.
23. B.S. C.P. 112:1971, The Structural Use of Timber.
24. *Lift Slab Slipform*, R. M. Douglas Ltd., Birmingham, England.
25. B.S.I. C.P. 110:1972, The Structural Use of Concrete, Part I.

26. H. D. Morgan, A Contribution to the Analysis of Stress in a Circular Tunnel, *Geotechnique*, Vol. 11, London, March 1961.

27. N. D. Pirrie, Design and Development of a Rock Tunnelling Machine, *Tunnels and Tunnelling*, March–April 1970.

28. J. Schimazek and H. Kratz, *Gluckauf*, Vol. 106, No. 6, p. 274, 1970.

29. Finns Drive World's Longest Rock Tunnel for Bargain Price, *World Water*, September 1978.

30. C. Muller, New Swiss Tunnel Lining System, *Neue Zürcher Zeitung, Technik*, No. 282, June 1971.

31. K. Angerer, Recent Tunnelling Experiences in the Alps, *Build International*, April 1968.

32. H. N. Steenson, Fast Set Shotcrete in Concrete Construction, *Journal of the American Concrete Institute*, June 1974.

33. The Victoria Line, Paper 7270S, *Proceedings of the Institution of Civil Engineers*, London, 1969.

34. W. Morse, Greenside-McAlpine Tunneller, *Contractor's Plant Review*, London, June 1969.

35. J. V. Bartlett, A. R. Biggart, and R. L. Triggs, Bentonite Tunnelling Machine, *Proceedings of the Institution of Civil Engineers*, London, Paper 7577, February 1973.

36. The Bentonite Tunnelling Machine, Nuttall Priestley, Trade Literature.

37. Mini Tunnels International Ltd., Trade Literature.

38. Jacking Concrete Pipes, Concrete Pipe Association of Great Britain, 1975.

39. B.S.I. Draft, C.P. For Falsework, 1975.

40. B.S.C.P. 2003:1959, Earthworks.

41. A. A. Lilley, Cement Stabilized Materials, *Concrete*, October 1974.

42. F. Schlosser and H. Vidal, *Reinforced Earth*, French Soil Mechanics Committee, March 1966.

43. A Total System of Slipformed Pavements, *Concrete*, April 1973.

44. *Sandwicks*, Cementation Ground Engineering Ltd., Rickmansworth, England.

45. B.S. C.P. 2004:1972, Foundations.

SUGGESTED READING

1. Caterpillar Performance Handwork, 9th ed., Caterpillar Tractor Co., Peoria, Illinois, 1978.
2. *Civil Engineering Procedure*, 3rd ed., The Institution of Civil Engineers, 1979.
3. R. Pilcher, *Appraisal and Control of Project Costs*, McGraw-Hill, New York, 1973.
4. J. A. Havers and F. W. Stubbs, Jr., *Handbook of Heavy Construction*, 2nd ed., McGraw-Hill, New York, 1971.
5. R. L. Peurifoy, *Construction Planning, Equipment, and Methods*, 3rd ed., McGraw-Hill, New York, 1979.
6. H. N. Ahuja, *Construction Performance Control by Networks*, Wiley-Interscience, New York, 1976.
7. *An Introduction to Engineering Economics*, The Institution of Civil Engineers, 1969.
8. H. N. Ahuja, *Successful Construction Cost Control*, Wiley-Interscience, New York, 1980.

INDEX